COPERNICUS

COPERNICUS

IVAN CROWE

TEMPUS

First published 2003

PUBLISHED IN THE UNITED KINGDOM BY:
Tempus Publishing Ltd
The Mill, Brimscombe Port
Stroud, Gloucestershire GL5 2QG

PUBLISHED IN THE UNITED STATES OF AMERICA BY:
Tempus Publishing Inc.
2 Cumberland Street
Charleston, SC 29401

© Ivan Crowe, 2003

The right of Ivan Crowe to be identified as the Author
of this work has been asserted by him in accordance with the
Copyrights, Designs and Patents Act 1988.

All rights reserved. No part of this book may be reprinted
or reproduced or utilised in any form or by any electronic,
mechanical or other means, now known or hereafter invented,
including photocopying and recording, or in any information
storage or retrieval system, without the permission in writing
from the Publishers.

British Library Cataloguing in Publication Data.
A catalogue record for this book is available from the British Library.

ISBN 0 7524 2553 6

Typesetting and origination by Tempus Publishing.
Printed in Great Britain by Midway Colour Print, Wiltshire.

CONTENTS

Acknowledgements 7
Introduction 9

1. Medieval Poland 13
2. Medieval Science and Technology 23
3. Legacies of the Ancient World 35
4. The Education of Nicholas Copernicus 41
5. Astronomical Instruments 49
6. The Bishop's Nephew 57
7. Medicine at Padua 65
8. A Renaissance Man 73
9. Questionning Old Beliefs 81
10. The Canon of Frauenburg 101
11. Astronomy 115
12. Contemporary Influences 123
13. Warfare 129
14. Friends and Fraternity 137
15. Copernicus the Astronomer 141
16. The Revolution of the Spheres 151
17. The Copernican Revolution 161

Glossary 177
Bibliography 182
A Note on the Illustrations 183
List of Illustrations 183
Index 185

ACKNOWLEDGEMENTS

This book would not have been written without the help of Sir Patrick Moore, the astronomer Dr Francisco Diego of University College London and the medievalist Professor David Abulafia of Gonville and Caius College, Cambridge, who each took the trouble to read the entire manuscript and evaluate the various accompanying illustrations and diagrams. Equally important was the advice offered by Professor Alan Chapman, whose knowledge of both the medieval world and astronomy provided a unique insight into the period, that enabled me to untangle fact from fiction. I must also thank Rob Warren, Robert Massey, Emily Winterburn and all the other staff of the Royal Greenwich Observatory for their time and help in supplying me with so much valuable information. Last but not least I wish to thank my family and friends for their support and in particular Omer Roucoux for his assistance in unravelling some of the more obscure questions that arose during the course of my research.

INTRODUCTION

Nicholas Copernicus is most famous as an astronomer. He lived during the medieval period: which we sometimes refer to as the Middle Ages. He had none of the sophisticated equipment we have today only his intelligence and willingness to question established beliefs, and his skills as one of the best mathematicians of his day. Thus he changed our ideas about the universe and the place our planet occupies in it. Because of Nicholas Copernicus we know that the earth revolves around the sun and that the universe is very much bigger than people ever before realised – and all this at a time when new ideas were not always welcome. In medieval Europe people's concepts of the universe and the place occupied by the stars and planets were almost solely informed by their religious beliefs and the Christian teachings of the Catholic Church, and you could have faced severe penalties or even execution just for saying the wrong thing. Many of the things Copernicus spent much of his life trying to prove we now take for granted but 500 years ago, in the world in which he lived, they were revolutionary.

But Nicholas Copernicus was not just an astronomer – he was many other things too. He was extremely well educated and a man of many wide-ranging talents who was highly respected for both his knowledge and experience of tackling some of the most difficult problems bothering the medieval society in which he lived: he was in fact the trouble-shooter of medieval Europe consulted even by popes. Consequently, he had as both his friends and protectors some of the most powerful people of his time and exciting and dangerous times they were too, when whole communities could be wiped out by either disease or warfare.

INTRODUCTION

The scholarly Copernicus himself was embroiled in politics on occasions and even briefly involved in armed conflicts and skirmishes that eventually helped reshape the map of Europe.

It was a time of change in other ways too, when many of the things that people had accepted for centuries without dispute were now being challenged and a few brave people were daring to question long established beliefs upon which their very society was based. It was period in which the power invested in the Inquisition by the Catholic Church sent many a dissenter to a grisly and horrific death but also an age when many of the new ideas that were to help form our modern world were first introduced. Innovations and inventions such as the printing press were destined to have a lasting influence on society, leading to an expansion of knowledge that played a fundamental role in the renaissance of science, the arts and medicine, as well as the religious reformation that was occurring in fifteenth- and sixteenth-century Europe. It was Nicholas Copernicus as a mathematician and an astronomer who was, in fact, to affect one of the greatest and most significant changes of all – not in any physical sense – but far more importantly in the way people thought.

MEDIEVAL EUROPE IN THE LATE FOURTEENTH CENTURY

1. The world Copernicus knew was largely shaped by events that had occurred during the century preceding his birth. The last of the Muslims were driven out of Spain and the city of Constantinople fell to the Turks. By the time Copernicus had begun writing the book that was to change people's view of the universe, the map of Europe and the surrounding regions had changed dramatically. One result of this was to be the introduction of new technologies and new ideas but also equally importantly the reintroduction of previously forgotten knowledge.

2. Nicholas Copernicus was born in 1473 at Torun – a city founded on the banks of the Vistula River by the Teutonic Knights, an order of warrior monks that had originally been invited into the region during the thirteenth century by one of the Polish rulers, Duke Konrad of Masovia, to help him suppress the local Prussian tribes who had begun to make raids into his domains. The Prussians were pagans and at the time, situated on the eastern borders of Poland, Torun was on the very frontier of Christendom.

I

MEDIEVAL POLAND

Nicholas Copernicus was born in 1473 at Torun – a city founded on the banks of the Vistula River by the Teutonic Knights – who were to be a recurring influence in the life of Copernicus. The Teutonic Order enjoyed the same standing as the Knights Templar and the Hospitallers of St John and like these two orders had similar privileges bestowed on them by the Church, one of which was that they were only answerable to the pope in Rome. Although originally founded during the Third Crusade, in Jerusalem, as a hospital order specifically dedicated to caring for ailing or wounded German knights and soldiery, the nature of their organisation very soon changed and several centuries before the time of Copernicus the Teutonic Knights had already essentially become a group of warrior monks whose main objectives were militaristic and self serving, rather than to serve as hospitallers. Even as hospitallers their partisan attitude had been noticed and commented upon by one crusader, John of Wurtzburg, who noted that anyone visiting their hospital, known as the 'German Building', who speaks any other language, 'can barely receive a blessing'.

Having eventually been driven from the Holy Land in 1291, when Acre fell to the Muslims, the Teutonic Order relocated themselves in Europe and sought lands and power by offering their services as a military force to various monarchs and princes. In 1211 the order was invited by King Andrew II of Hungary to protect his borders against the Transylvanian Cumans by colonizing their land and converting these pagans to Christianity. The

3. Medieval warfare was brutal, as was the treatment of captives unless they were wealthy and there was a chance of obtaining a ransom. Nevertheless there were no guarantees. Many a lord was slain, mutilated or incarcerated for life to settle a grievance. The sacking of a town by a medieval army preceded by the call 'cry havoc' was also something to be feared and dreaded. The local countryside fared no better from a marauding army looking for food and seeking booty. Robbery, rape and pillage were the order of the day until the battle had been decided. For this reason soldiers who fell into the hands of their enemies, or in particular the local population, were often tortured to death as a matter of course as repayment for the atrocities they themselves had perpetrated.

Teutonic Knights were given extensive rights of autonomy in exchange for their services; but their demands became so excessive that they soon found themselves expelled from Hungary. The Teutonic Order did not let this temporary reverse stand in the way of their military and territorial ambitions and before long they had yet another opportunity to carve out a new country for themselves on the outer fringes of medieval Europe in a region which is now part of Poland, by conquering the non-Christian Prussians who then lived there, whom they considered heathens ready for conversion by force of arms. At that time it was quite commonplace for people to treat those who did not share their own beliefs as enemies, even to the extent of going to war with them. Indeed, we still find some people attempting to do the same today and similarly to today they very often had a hidden agenda when they decided to engage in such conflicts, such as to gain power, influence or wealth. Despite the Teutonic Knights' Christian vows of poverty, chastity and obedience, their objectives appear to have been no different to any other medieval army's and seem to have done nothing to diminish their desire for worldly goods nor their brutality, in an age when such things were common and still seen as an inevitable consequence of warfare, if not an acceptable norm.

Their presence in Polish territory had been one of those cases of history repeating itself. The Teutonic Knights had originally been invited into the region by one of the Polish rulers, Duke Konrad of Masovia, during the thirteenth century when the local Prussian tribes began to make raids into his domains. He had asked the knights for help in quelling the Prussians to deter them from continuing their incursions. The order of knights had accepted the challenge and proceeded to prosecute the war with the accustomed ferocity that is normally associated with warfare during that period. In fact the conflict was if anything more brutal than usual, with no quarter being given on either side and no prisoners being taken. That is apart from those that were to be executed and executions were often carried out in various sadistic ways calculated to maximise the prisoners' suffering and discourage further opposition. The Prussians were not Christians but pagans and the Teutonic Knights had no regard for the lives of a pagan people, believing that non-Christians were of no worth and thus did not need to be accorded any mercy. The Prussian tribes were equally merciless, inhumane and brutal in defending their territories and treated their Christian captives in much the same way.

Initially the armies of the Teutonic Knights had the upper-hand due to their superior equipment and more advanced armament. The Teutonic Knights had siege engines for example with which to bring down the fortifications of Prussian towns and castles. Moreover, the infantry deployed by the Teutonic Knights had crossbows that could propel an iron bolt far further than any weapons the Prussians initially had available. In fact at first they probably had little more than their normal bows and arrows or spears with which to retaliate. But they learnt quickly and the Prussian tribes were soon able to produce and arm themselves with much the same kinds of weapons as their opponents. In fact each time the Teutonic Knights acquired some advantage the Prussians emulated them, or were able to respond with an effective counter-measure. Thus the war became an arms race with each side trying to outdo the other. In consequence the conflict dragged on for decades and was only finally brought to a close when most of the Prussians had either been annihilated or had been persuaded, by the liberal use of torture and mass executions, to convert to Christianity, or had at least pretended to do so. Men, women and children had been slaughtered with equal disdain and the same total disregard for human life. It was in reality nearly complete genocide: a medieval form of ethnic cleansing that was carried out with the same ruthlessness and bore the same hallmarks of barbarity that we associate with such wholesale massacres today.

Once the war was won and his lands secure, the Polish duke, Konrad, intended to annex the Prussian territories. But with the Prussians finally conquered and their lands occupied, the Teutonic Order did not want to leave or relinquish their hold over the land. This was hardly surprising. The war had been hard fought and in those days the booty of war was always part of the reward for the victor, and in this case there was an added bonus: the newly occupied territory over which the Teutonic Knights wished to impose their own rule. They not only used various tricks to delay their departure, moreover they invited others to join them. Many additional colonists came from Germany as had the knights themselves.

Duke Konrad would not have minded the knights staying and settling in Prussia if they had accepted his suzerainty as their overlord; but this was not enough for the Teutonic Order who wished to have complete sovereignty over the country. Despite all the efforts of Duke Konrad and his successors they could not be shifted. Further battles were fought and even the pope was asked to intercede but by a mixture of force and subterfuge the Teutonic Order nevertheless kept the land they had won and held it until just a few decades before Copernicus was born, when a royal marriage between the king of Poland and the princess of Lithuania united these two neighbouring countries who then, together, took up arms against the Teutonic Order.

Until the marriage the Lithuanians had been pagans too and although true conversion may have been largely confined to the aristocracy falling in line with their new monarchy, this did not apparently diminish their effectiveness or commitment as allies. With the joint might of these two countries arrayed against them the outcome for the knights was probably inevitable; but the knights did not stand alone. Although only a small force of a few hundred men themselves, they were supported by their vassals and accompanying men-at-arms. They were wealthy as an Order and their forces were thus supplemented too by the usual mercenaries attached to medieval armies, as well as various adventurers who came from all over Europe both to obtain experience of warfare and in search of plunder.

Fighting on the side of the Teutonic Knights, an order of hospitallers, also technically gave them holy war privileges exempting them from retribution in this world and the next for any excesses perpetrated during the conflict, even though in this case they were fighting other Christians. This was a point not missed by some who were later to take their complaints direct to the Vatican in Rome.

Many of the mercenaries would have been Milanese crossbowmen and Swiss pikemen, as well as English longbowmen and German heavy cavalry. They also had powerful allies. One was the king of Bohemia, another was Charles of Burgundy and a further notable among the adventurers was Henry Bolingbroke, the future King Henry IV of England who was eventually to found the Lancastrian dynasty. However, despite this foreign support the combined forces of Poland and Lithuania successfully defeated the Teutonic Knights, finally destroying their army and their power in 1410 at the battle of Tannenberg. The result was that the lands that the Teutonic Knights had controlled from then on became vassal states of Poland. The knights lost much of their possessions including the city of Torun which they had originally founded as an outpost to guard the borders of their territory. But they were allowed to keep one region – the duchy of Ermland – where they were still free to exert an oppressive control over the unfortunate, by now much reduced and subordinate, native Prussians. Nevertheless, for centuries afterwards there was a mixture of conflicts, rebellion and uneasy alliances between the rulers of Poland and either the Teutonic Knights, or their Prussian subjects and as we shall see later – even a renewal of outright warfare. The continuing resonance of these events was to have a significant influence on the life of Nicholas Copernicus and his family in a number of ways: some subtle but others far more direct and confrontational.

At this time Poland was essentially a frontier on the very edge of Christian civilisation. Like all new frontier countries, it was perceived by many as virgin territory, offering opportunities not always so easily available in longer, well established communities, in which some people may have had a vested interest in maintaining the status quo. The possibilities for trade and personal advancement in the region attracted immigrants from far and wide who hoped to make their fortune there. People came from all over Europe and even parts of Asia. It was an important port for trade with Russia and other non-Christian territories on the eastern borders of Europe as well as providing trading links with merchant travellers acquiring more exotic goods and commodities from the Middle East and the Orient in general. The population of Poland was therefore very cosmopolitan with a mix of people from many different backgrounds, countries of origin and religious groups. This in itself would have been unusual within the normal bounds of medieval Christian Europe. The majority of the population were Catholics, but there were also many living there because they were not, and they found it easier in a more remote part of Europe to avoid the strict rules laid down for Catholics by the Church of Rome, whose established teaching and authority they may have been seen to challenge. And there were Jews too: many seeking to escape the persecution they had experienced in Germany and even including some from Spain following the rise of the Inquisition, as well as those driven out of several other countries in Europe that were not given to religious tolerance.

All immigrants, including persecuted minorities, were welcome and were viewed by those in power as potential contributors to the economic growth of Poland. Apart from anything else many of the immigrants coming from other parts of Europe, or sometimes quite different cultures, could offer new skills or provide valuable expertise, for example as craftsmen, merchants or even maybe as financiers. It is certain that some may have been impoverished when they arrived but others would have been well heeled if not actually wealthy, bringing with them money acquired in their former homelands, that could encourage trade and help power the dynamo, effectively promoting further commerce. This attitude was in part inherent, as even in pagan times the original population had generally accepted newcomers from foreign parts in much the same enterprising spirit and furthermore, had not shown the same intolerance towards other faiths as the Church of Rome did.

One of the great attractions that must have drawn many immigrants to the region was the presence of the Hanseatic League and the advantages it offered for tradesmen and tradeswomen, merchants and would-be entrepreneurs, to whom towns like Torun would have particularly appealed as a base for their business. The Hanseatic League was originally founded by north German towns and German merchant communities abroad. The word 'Hanse' actually means guild or association and like the other guilds associated with medieval England and elsewhere in Europe, it represented a federation of tradespeople dedicated to promoting and protecting their mutual interests. The Teutonic Order had played a very active part in the process of promoting German interests in both Europe and elsewhere by maintaining a powerbase in far-flung places to protect German merchants abroad and establishing new towns throughout Prussia before they lost their control of the region after their defeat at the battle of Tannenberg.

Following this reverse of fortunes, many of the towns founded by the Teutonic Knights were lost to them and although one of these was the towns of Torun, the Hanseatic League survived this change and under Polish rule trade continued as before. With their far-flung trading connections established in the thirteenth century the organization had by now become one of the most dominant merchant groups in Northern Europe. Among other things they were one of the main suppliers of exotic goods, which the League's members often acquired via a series of trade links with other merchants who sometimes

4. In medieval Europe all Christians were forbidden to profit from usury, which meant they were not allowed to profit by charging any interest when lending money. There was therefore little incentive to persuade Christians to take the risk inherent in extending a loan or giving credit. However, there was no such prohibition on Jews and thus many Jews made their living as moneylenders. Those borrowing money may have on occasions complained about the repayments they had to make; but the fact that people could borrow money probably helped encourage those of an entrepreneurial spirit, which meant that both trade and industry often flourished as a result.

5. The house where Copernicus lived.

had access to merchandise from the most distant of foreign lands. Even goods from India and China reached Europe through this network and they had a near-monopoly of the long-distance trade in the Baltic. This opened up trade to the East; and whereas the merchants of German towns in the Rhineland, like Cologne, were trading with England and Flanders, where Bruges played a major role, it was the merchants of Poland and Prussia, in towns such as Torun, that maintained trade links with Russia through centers such as Novgorod and the eastern Baltic, where the Hanseatic League helped establish the ports of Riga and Danzig.

Among the many immigrants that came seeking a new life in Poland was the family of Nicholas's father, who probably originated from a region known as Silesia, which is now part of Germany. In fact this still remains a source of contention among scholars in both Poland and Germany as regards Copernicus's true nationality with both countries claiming him as their own. Originally the family name was probably Koppenigk. The father of Nicholas Koppenigk whom we know as the astronomer Copernicus was also called Nicholas. It is known that he became a merchant and was among other things a dealer in copper, which he imported from Hungary. The name Koppenigk actually originally meant someone who deals or works in metal and particularly copper. Therefore it was probably a trade with which the family was

6. Sacks and bales surround this medieval merchant as he examines his ledgers. There even appears to be a millstone standing among the other goods awaiting dispatch. Writing instruments and a form of abacus or tally lie on his desk. At the time of Copernicus Poland was essentially a frontier on the very edge of Christian civilisation and the possibilities for trade and personal advancement in the region therefore attracted immigrants from far and wide who hoped to make their fortune.

long associated. It was only much later that we find it changed into the Latin form of Copernicus. This was possibly done by the younger Nicholas himself when he was a student. This was a commonplace practise among both students and academics of the day, as Latin at that time was the language of learning throughout Europe, used by churchmen and academics alike.

Nicholas's father lived for a while in Kraków, which at that time was the capital city of Poland, but later moved to Torun. He must have been successful in his trade and was obviously well respected there because he became a leading citizen, as well as a magistrate of the city. It was here too that he met and married Nicholas's mother, Barbara Waczenrode. Her family, whose ancestors themselves had been immigrants several generations before and were probably also of Germanic origin, were well connected and very wealthy. This was later to give the yet unborn Nicholas special advantages which provided him with powerful family ties that were destined to play such an important part in shaping his future.

7. When Johann Gutenburg produced the first printed Bible at Mainz, in Germany, the young Nicholas Copernicus was only nine years old. Printing had been used in China and also occasionally in Europe for centuries using woodcuts. The new idea Gutenburg introduced was moveable type, with each letter of the alphabet being made separately as metal type that could be used repeatedly rather than just once. The art of printing was a great aid to both communication and learning. Printing meant that the same information could be reproduced cheaply and quickly over and over again. Before this, all copying had to be done by hand – usually by monks and consequently at the time relatively few people outside the Church could read or write. Inventions such as the printing press were destined to have a lasting influence on society, leading to an expansion of knowledge that played a fundamental role in the renaissance of the arts, science and medicine, as well as the religious reformation that occurred in fifteenth- and sixteenth-century Europe.

II

MEDIEVAL SCIENCE AND TECHNOLOGY

The world in which Nicholas Copernicus grew up was one in which the printing press had only just been invented by Johann Gutenburg. When the first printed Bible was produced at Mainz, in Germany, the young Nicholas was only nine years old. Printing had been used in China for centuries using woodcuts – mainly to reproduce pictures. Woodcuts were also occasionally used in medieval Europe in a similar way; but the new idea Gutenburg introduced was a moveable type, with each letter of the alphabet being made separately as a metal type that could be used again and again. Woodcuts were still used too but only to illustrate the text. The art of printing was a great aid to both communication and learning. Printing meant that the same information could be reproduced cheaply and quickly over and over again. Before this, all copying had to be done by hand – usually by monks as at that time few people outside the Church could read or write.

This took a huge amount of time and made books very expensive indeed, well beyond the reach of ordinary people. The printing press changed this too by making books cheaper. Because more people could afford to buy them more of the population were encouraged to learn to read and write. Reading before was a requisite skill acquired by priest and other clergy, but many ordinary people could not read and even the those who were wealthy were sometimes illiterate and this occasionally included members of the aristocracy as well. One result was that even the ordinary person in the street had the opportunity to become more educated. It also stimulated new thinking and the exchange of ideas broadened people's general knowledge and allowed them to have more informed opinions on a wider range of topics.

This alone contributed a great deal towards changing medieval society and the introduction of printing using moveable type represents a crucial point in time and probably more than anything else is one of the key, pivotal events marking the cultural transformation we refer to as the Renaissance.

Another innovation also helped make books cheaper and that was the earlier introduction of paper-making in the thirteenth century. Paper is something we take for granted nowadays but it was originally unknown in medieval Europe. This art too came from China thousands of miles to the East. Previously all books or any other writing was on vellum or parchment. This was produced from animal skins. Like hand copying, this was also time-consuming and expensive and tended to make books rather thick and heavy.

The process by which paper-making reached Europe was a curious one: the Arab traders had been visiting China for a long time and had learnt from the Chinese how to make paper. During the Middle Ages the Muslim Arabs had crossed the Mediterranean Sea to attack and conquer Spain. Spain had remained under Muslim rule for a very long time. Although this was naturally resented by the Spanish people there had been benefits to Europe as a whole. There was a remarkable amount of religious tolerance under Muslim rule, with Jews, Christians and Muslims living and working side by side. Many new ideas had arisen as a result of the exchange of knowledge that occurred between the cultures. One result of this was that the art of paper-making had been introduced into Spain – and from there to the rest of Europe. This combination of a cheap and rapid printing process, along with the replacement of parchment or vellum by paper, increased the spread of knowledge and created a literate, far better educated population in medieval Europe.

Science, as we know it, was only just beginning to develop in Europe and hearsay or opinions based upon religious beliefs or superstition were still generally the rule in many subjects, rather than genuine research. That is not to say that learning by observation had not been attempted by some scientists and other great thinkers over the centuries, who had sought the truth about the nature of the world in which they lived, but there was nothing like the technology that we enjoy today that they could employ to help them. There were for example no telescopes available, or any of the scientific optical instruments that we are now familiar with, like modern microscopes. The art of glass-making had been well known to far more ancient cultures such as those of Egypt and the Romans, but had gone into decline with the fall of these empires. Even glass in windows was still something of an expensive rarity during

8. The mill wheels turn, activating the trip hammers that mash up the pulp required to make paper. In the next stage, a sieve is dipped into the pulp by the papermaker to form a thin layer that coats the mesh. Once this residue has sufficiently dried out, it is placed in a press in a stack with other sheets, separated by cloths, to compress it and expel the last of the water to finally produce paper.

Papermaking was introduced into Europe by the Arabs of Muslim Spain, who learnt the art from the Chinese with whom they traded. Producing printed books using velum made books too expensive for many people and vellum production could not have kept up with the vast increase in the number of books being produced. Paper was also more absorbent and enabled faster drying inks to be developed that increased printing speeds and allowed books to be manufactured more quickly and cheaply.

much of the Middle Ages in many parts of Europe and could normally only be afforded by those who were fairly wealthy.

Spectacles, though, had been known in Europe since at least the late thirteenth century. The first recorded reference to spectacles was in a sermon made by one Giordano di Rivalto, an Italian friar who demonstrated their function at a conference in 1285. He drew attention to the fact that they had by then been in use for barely twenty years, and so we can reasonably assume that spectacles were invented some time before that. The important question to ask here is by whom were they first made? The answer to this question is not clear and the mystery surrounding their invention and their first appearance on the public stage, tells us something significant about the Middle Ages and the attitude of the Church towards new scientific developments and technological innovation.

It is worth noting that when referring to spectacles at the conference Friar Rivalto says that he had spoken to the man who first made them while making it quite clear that the actual inventor, who was yet another reticent individual, wanted to remain anonymous. The first person he was talking of was probably Alessandro della Spina, as in the archives of the Monastery of Florence where he once lived it is mentioned that, 'He made spectacles himself that were first made by someone who would not divulge the secret of their manufacture'. The second of the two unnamed producers of spectacles is not so easy to identify though, but the reason for the secrecy about the person responsible for such a useful invention as spectacles may be far less difficult to define, given the suspicion with which new technology was viewed and the inherent dangers the author of such an innovation might face.

Who really invented them is not known for certain, but it seems highly likely that the English friar and innovator Roger Bacon may have been partially responsible. He made his first reference to his experiments with lenses in 1262 and had written, 'how useful this glass must be for those who are old or have weak eyes'. He had also told a friend of his, Heindrich Goethal, of his experiments with lenses to aid people with impaired vision. In fact Heindrich Goethal was actually asked by Bacon to visit the pope, Martin IV, to tell him of this invention but when Heindrich arrived in Florence he heard that the pope had died. This was in 1285. While in the city Heindrich had met the Friar Alessandro della Spina and told him of Roger Bacon's invention. Spina did not keep the secret to himself. Instead, he repeated it to another friend Salvinus de Armatus, who then began producing spectacles for the public himself: he introduced the idea as his own.

In fact by some it was he who was assumed to be the inventor of spectacles, because this we are told by one narrator, Peter Von Muschenbroeck, was stated on his gravestone. The inscription read as follows. 'here lies Salvinus de Armatus of the Armati of Florence, inventor of spectacles. God pardon him his sins'. The second part of the inscription gives some hint of the suspicion and censure such an accomplishment might attract from the more conservative elements of society and the Church. What we must remember when discussing the technological developments that occurred during the Middle Ages is that it was a period encompassing several centuries in which a scientist or an inventor who acquired new knowledge, understanding, or control of natural phenomenon, or simply succeeded in achieving some unforeseen

9. This drawing shows how prior to the invention of the printing press all manuscripts were hand-written, usually by clerics. This was time-consuming and thus made access to the 'written word' very elitist and expensive.

10. Spectacles had been known in Europe since at least the late thirteenth century. The first recorded reference to spectacles was in a sermon made by one Giordano di Rivalto, an Italian friar who demonstrated their function at a conference in 1285. He drew attention to the fact that they had by then been in use for barely twenty years and so we can reasonably assume that spectacles were invented some time before that but it is not certain by whom they were first made.

breakthrough, could so easily find themselves being accused of sorcery, to be then publicly beheaded or even burnt at the stake.

In fact Roger Bacon had attempted to remove some of the constraints the Church placed upon experimentation. Sometime, around about 1266 he had applied for papal permission to write a book extolling the advantages of experimental methods. The project was approved but he was told to send the book to the papacy secretly. Even popes were sometimes hesitant about supporting such endeavours. Bacon was not meant to tell his superiors anything of the book but by the time he had completed the work, the pope in question had died. This single event probably delayed the inclusion of experimental science into the curriculum of universities and a step that could have promoted the advancement of science was never taken. Bacon was imprisoned in 1277 for what his order, the Franciscans, termed 'novelties' in his teaching. He remained in prison for two years. This did not dampen his enthusiasm but he was an exception: most people receiving such an admonishment would have curtailed their activities and their tongues for fear of more brutal punishments.

Even the indomitable friar took steps to avoid trouble when he could. Roger Bacon had approached one pope to ask permission simply to write a book about the value of experimental science and furthermore had sent a friend as an emissary to a second pope to tell him of his new invention before attempting to make it public. Moreover, when a third party began producing spectacles he was in no hurry to claim responsibility for their original concept or manufacture. One reason may have been that a charge of sorcery might have followed for despite the praise accorded the unidentified inventor by Friar Rivalto, spectacles were nevertheless still condemned by some factions within the Church.

It seems ridiculous now but during the Middle Ages some members of the Church even criticised the idea of people using forks to eat instead of their fingers. They objected on the basis that it was an insult to God to try and improve upon the body he had provided and the natural fork – our fingers. The man or woman who tried to improve sight was going to be regarded with even more suspicion. This is not exactly a novel idea, a previous author, Gordon, drew the same conclusions. In his book, *Medieval and Renaissance Science*, he says,

> The natural philosopher who was bold enough to present to his prince a pair of spectacles or a telescope would be in imminent danger of being regarded by the Church as a powerful and dangerous magician of satanic origin.

It is these aspects of the age in which Copernicus lived that we have to bear mind if we are to understand the constraints and dilemmas that any man or woman with an enquiring mind, such as he and Roger Bacon faced.

When considering the story of how and by whom spectacles may have been invented, one further tantalising possibility comes to mind. If the development of such a useful aid to normal vision was shrouded in such secrecy; what else might have been surreptitiously invented in medieval Europe? For example, our Franciscan Friar Roger Bacon is also quoted as saying,

> Glasses may be formed that the most remote objects may appear just at hand and the contrary so that we may read the remotest letters at an incredible distance and may number things though ever so small.

It would appear on this basis that Bacon may have already been on the way towards at least recognising the possibility of using lenses to also construct a device, like a telescope for viewing distant objects and may have also considered producing a form of microscope, long before such inventions were to become publicly proclaimed by other people centuries later. He certainly seems to have had the knowledge and insight to see the potential offered by optical lenses and had experimented with them even if he never built a permanent instrument. How many discoveries were made and never saw the light of day due to the political and religious climate we shall never know. That it happened on occasions seems highly likely and one can only speculate on the advantages that might have been gained had this not been the case.

11. Pope Gregory IX thought that anyone who disputed the authority of the Catholic Church should die and it was he as pope who instigated the Inquisition that was responsible for many barbaric acts and sent so many people to their deaths. He died in 1241 aged ninety-nine.

A Dutch eyeglass maker, Hans Lippershey, is usually given the credit for inventing the telescope in 1608, long after Copernicus's death. This is the date on which he offered it to the government for military use. Although he is generally recognised as the inventor of the telescope, it is said that he only made his discovery as a result of children playing with lenses in his workshop. It has been claimed that it was they who discovered that a church tower seemed to become nearer when they held up two lenses together – one concave and one convex – in line with their eyes; and that Hans Lippershey merely took advantage of the children's observations to take credit for the invention himself.

Regardless of whether or not this story is true, there seems no reason why something like this could not have happened several times before – and if so why was the idea of remote viewing never developed until the seventeenth century nearly 350 years after Bacon first made his perceptive observations about the properties of lenses? The answer may simply be fear. The fear of being accused of sorcery or witchcraft may have held back the advancement of not only optics but also other sciences too. An individual who discovered that they could bring distant objects closer or increase their size simply by looking through one or two pieces of glass would quite naturally have been terrified of being seen as dabbling in the black arts. Indeed, with no other explanation, even they themselves may have believed it to be magic. Consequently the vast majority of people who witnessed such phenomena would have probably simply kept quiet.

We have good reason for thinking that this was the case. Whether anyone ever did construct a telescope and use it before Hans Lippershey in 1608 we shall probably never know. Nevertheless, that there was some secret activity regarding the science of optics now seems beyond doubt: for that we can largely thank not a scientist or a historian but an artist. The observations of the artist, David Hockney in his recent book *Secret Knowledge* raises some interesting questions that go well beyond the immediate remit of the book. In *Secret Knowledge* Hockney discusses the possible use of optical devices by some medieval and renaissance painters. This is something of which there is no official record; but the insights that David Hockney as a painter brings to this research makes his conclusions seem irrefutable.

Essentially what he noticed was that after a particular date, around 1430, the paintings and drawings of certain artists seem to change. The images produced by some artists were practically photographic. He had already become interested in the work of a later painter, Ingres, who he suspected had used a device known as a *camera lucida* to achieve a better likeness in his portraits. The *camera lucida* is basically a prism that allows the artist to see both the subject he is illustrating and the surface he is working on at the same time. Hockney had also made some experiments himself to see if it helped him obtain similar results. This alone had made Hockney more aware of the elements to look for in a picture produced using an optical aid, one of which is the strong lighting that is important to illuminate the subject and another the heavy shadows this produces. These were the very characteristics he had found evident in many pieces of art work, but some of the pictures he examined had been drawn or painted several centuries earlier than those he had been originally studying. Some had even been produced during the Middle Ages, long before suitable optical devices were thought to have existed.

He was intrigued and embarked on a course of personal research to find the answer to the mystery. It had already been suspected by some art historians that artists like Canaletto had employed a *camera obscura* to improve their work and it may be the reason why Canaletto demonstrates such a good understanding of perspective in his paintings. But what David Hockney had discovered suggested that a range of other methods were being used too, involving alternative optical devices that employed both mirrors and lenses to create a more accurate image of the artist's subject. It also gradually became apparent to Hockney that not only were these practices probably widespread in medieval Europe but they also appear to have been generally employed surreptitiously. The reason why such useful developments in art were carried out so secretly may once again have been fear. Hockney himself makes much the same comment, suggesting that, 'those who revealed the 'secrets' of God's kingdom might be accused of sorcery and burnt at the stake'. Such fear must have suppressed technological advances in medieval Europe. Modern historians for example knew nothing of these

12. To set up a sundial the *gnomon*, or pointer, has to be aligned with the earth's axis to ensure that it compensates for the 23.5 degree angle that the earth is inclined to the plane of the elliptic, which corresponds to the imaginary line that the sun appears to follow as it crosses the sky. Although the sundial is only used during the day, the alignment would often be done at night by lining up the *gnomon* with *Polaris,* the Pole Star. This meant that the angle of the *gnomon* would have to be progressively decreased in relationship to the base plate at lower latitudes the further south one went. For this reason, the number of degrees that the *gnomon* is tilted is exactly the same as the degree of latitude. This means that if you knew the precise degree of latitude you could also adjust the sundial by simply changing the angle. Some portable sundials were in fact adjustable in this way.

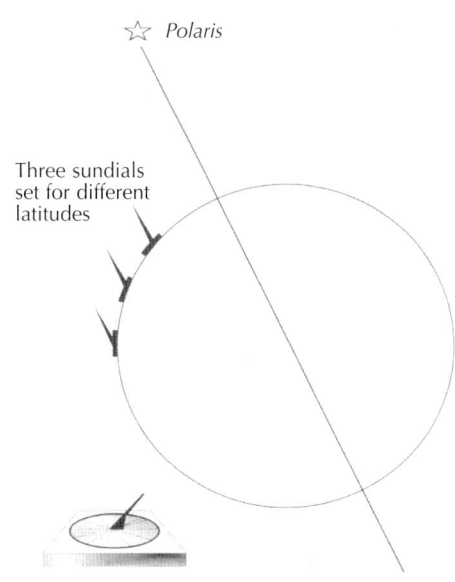

techniques. They were never in the public domain, nor overtly recorded and thus all knowledge of them was eventually lost. One is tempted to ask how many other scientific developments were similarly stifled at birth.

There was also a lack of accurate, mechanically powered timekeepers in Copernicus's time. The few clocks that existed were normally only found in buildings like churches or other public buildings and were run using a system of weights and pulleys. You were lucky if they were correct to the nearest hour! Other methods used to mark the passage of time were water clocks, that relied upon measuring the amount of water that was left in a container with a set outlet; using an hour glass: that worked like a glass egg-timer, or a burning candle with marks to indicate how long it had been burning. The last of these in particular was not very reliable. But the limitations of these timepieces were of far less importance to the medieval astronomer than one might imagine.

The most dependable form of timekeeper available during the Middle Ages was the heavens. Firstly there were sundials that used the movement of the sun's shadow to indicate the time of day. The most familiar form of sundial and probably the most common one in use during the Middle Ages, was the kind one still occasionally sees today in some gardens. These had to be sighted very carefully if they were to accurately mark the passage of time. The sundial had to be correctly orientated and the *gnomon*, which is the rod that casts the shadow on the normal horizontal-based, or wall-mounted perpendicular-fixed sundial, had to be angled so that it was aimed in the direction of the Pole Star

The night sky when facing North

13. The earth rotates around an axis in line with the Pole Star, *Polaris*. In the Northern Hemisphere, this makes the stars appear to move from east to west as the earth turns: so that they seem to revolve around *Polaris* during the course of the night. *Polaris* is the star seen in the centre of the circle. Although the earth turns once every twenty-four hours, the additional movement of the earth as it orbits the sun means that the entire night sky will actually appear to complete a revolution in just under twenty-four hours: twenty-three hours, fifty-six minutes and four seconds to be exact. This results in the position of the stars being just a shade less than four minutes earlier each successive night throughout the year.

Ancient people, who learnt to take account of this small discrepancy in the position of the stars each night, were able to tell the time at night by referring to this giant celestial clock. During the Middle Ages people in Europe and the Islamic world often used a device known as a nocturnal to help them find the time after dark.

The four illustrations to the left show how far the stars appear to move over a period of time: in this case a duration of three hours. The earth turns a full 360 degrees every twenty-four hours. Therefore even allowing for the four mintues difference each successive night, this represents approximately fifteen degrees of rotation every hour. One can more easily detect this movement by carefully observing how a particular constellation of stars such as the Plough is affected.

It is the movement of the constellation called the Plough, sometimes known as *Ursa Major* or the Great Bear around *Polaris*, that has been used as an example here in the smaller illustrations shown to the right of each star map, because this was one of the constellations that was most commonly used as a star reference when employing a nocturnal in medieval Europe; the other was *Ursa Minor*.

14. The revolution of the earth makes all the other stars in the sky appear to revolve around the Pole Star. During the Middle Ages it was thought that it was actually the stars moving, but even so this phenomenon meant that people could tell the time at night by using an instrument called a nocturnal. This device consisted of two circular plates and a pointer. The top plate first had to be moved until the indicator lined up with the correct date of the calendar inscribed on the rim of the bottom plate: this is because the stars are not in exactly the same position every night. Then the nocturnal was held upright and the hole in the centre lined up with the Pole Star, so that it was visible to the observer. Once this was done the observer moved the pointer until it was in line with two stars *Dubhe* and *Merak*, indicated in the constellation of *Ursa Major*, the Great Bear. This would bring the pointer in line with the correct time as displayed on the edge of the top plate.

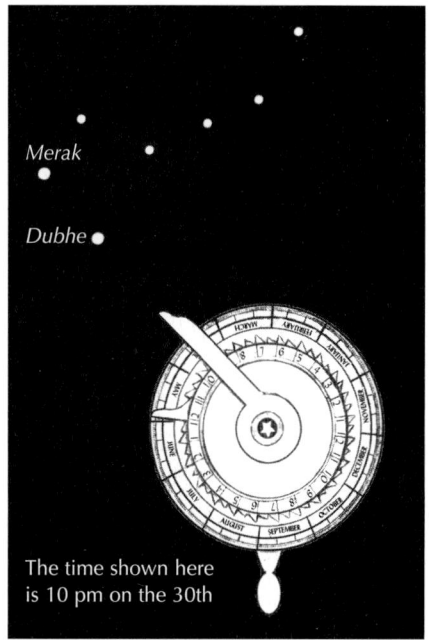

The time shown here is 10 pm on the 30th

to ensure that it exactly lined up with the earth's axis. The fixed settings for this type of sundial had to be adapted to suit different latitudes, but there were several other designs of sundial that could be adjusted to work at any location.

Although sundials could only be used during the day, the medieval astronomer had another less well-known device called a nocturnal that could be used to also tell the time at night. This consisted of a circular plate with a hole in the centre and a moveable arm. The user had to know his or her location in terms of latitude and also the date: the instrument was marked with calendar settings for making the appropriate adjustments. Once this was done the nocturnal had to be held so that the Pole Star could be seen through the central hole and the observer moved the arm until it lined up with the two pointer stars in the constellation of *Ursa Major*, the Great Bear. The alignment of the arm against the measurements on the dial of the nocturnal would give the time.

In fact these two methods of telling the time were so much more accurate that all other alternative timekeepers, including the early clocks were checked and calibrated against them. Many clocks actually came with a sundial supplied for this purpose so they could be corrected – for example at noon each day. Therefore, even though Copernicus when he began his studies and research would probably have had to face and overcome many challenges that a modern astronomer would never meet, being able to mark the passage of time would not have necessarily been one of them.

15. The attack of Constaninople by the Turks in 1453.

III

LEGACIES OF
THE ANCIENT WORLD

The Roman Empire at its zenith had conquered both Egypt and Greece and had absorbed much of the knowledge, learning and even culture of these two civilisations. When eventually Rome itself was overrun by the barbarian Goths and Vandals only the easternmost part of its once vast empire survived relatively untouched. But gradually its power declined and the Eastern Roman Empire contracted until and the city of Constantinople, which we know as Istanbul, became the last outpost of the once powerful Roman Empire as well as the guardian of the accumulated knowledge of three civilisations.

In 1453 the city of Constantinople fell to Turkish invaders and the vast majority of the Christian population fled to Europe to escape Islamic rule. Some of these people brought with them copies of manuscripts from ancient Greece. The common language of educated people in this eastern part of the old Roman Empire had been Greek, which was once even spoken in Egypt and particularly in the city of Alexandria where several important astronomers had lived in the past. Many of the manuscripts were therefore in Greek, a language that was also understood by some scholars in Europe, and among these were treatises on medicine, mathematics and the writings of early astronomers such as Archimedes and Ptolemy.

The city of Rome, as the seat of the pope was once again powerful as the Vatican excercised its influence over the Catholic Church. Much of what was taught by the Church and most academics came from the writings of early Greek and Roman teachers and philosophers. A philosopher is someone who seeks the

16. The occupation of Spain may have been regarded as a tragedy by some in the Christian world, but it led to the introduction of Arabic numerals and more advanced forms of mathematics and geometry into Europe, providing the tools by which scientists could achieve a better understanding of our universe and the world in which we live. There was also a remarkable amount of tolerance shown towards other faiths under Muslim rule. There were Jews as well as Christians and Muslims living and working together in Spain during this period and a great deal of mutually beneficial, cross-cultural exchange occurred as a result of this. Cordova was probably one of the most highly regarded centres of learning and education in the medieval world and much of what was taught there filtered through to Europe.

truth, but most of these philosophers had lived and died centuries before and although much of what they had said was correct a significant amount was simply speculation or theory and quite often wrong. The knowledge of these ancient civilisations often was genuinely superior but in the Middle Ages there was nevertheless sometimes undue reverence for the teachings of the ancient world and as a consequence few questioned the wisdom of the past. In fact frequently, it was quite literally considered to be an article of faith for Christians to accept some of these ideas without question. The Church frequently supported this point of view when past ideas supported its own teachings.

One of the legacies of the Ancient World was roman numerals. Everyone is familiar with them because we occasionally still use them today. Many clocks, such as those on church towers or public buildings and some watches, have Roman numerals instead of the modern digits. They are often used when printing the date of a book's publication and you will for example sometimes see them at the end of a film or television programme telling you the date it was made. Roman numerals were still in use in Europe during the early part of the Middle Ages and were for a time the only form of numbers being used in most of the Christian world and this could create difficulties.

The problem with Roman numerals was that they made reliable calculations involving large numbers very difficult – if not at times impossible. In modern mathematics all the numbers, from a single unit to huge numbers using millions or even trillions, are written using the same set of figures from 0 to 9, whereas in Roman numerals 1000, 500, 100, 50, 10, 5 and units less than 5 all have separate symbols. To make it more complicated a lower number to the right is an addition and a lower number to the left means it should be subtracted from the bigger number to find the value. The number 9 is one example of a lower number being subtracted from a higher number i.e. 10 – 1 = 9 and is expressed as IX. Other simple examples would be 4 as IV and 90 as XC, but when you get to larger numbers far more complex sequences are involved. Even writing the date can appear complicated and can also be relatively difficult to decipher for some people, particularly when a simple four-figure number like 1946 becomes MCMXLVI.

There were short-hand methods for writing very large numbers. A number like 36,000 could be written as thirty-six followed by a dot and the Roman symbol for 1,000 – XXXVI.M for example. Problems arise though when trying to add two or more exceptionally large numbers together. In some cases it could take a considerable amount of time even for those who knew how – moreover the risk of an error creeping in would be very high indeed. The problems that could occur with more complicated calculations mean that it is not even worth contemplating the attempt and this placed a severe limitation on the development of mathematics in Europe for centuries. It also placed a severe limitation on the recording of scientific observations and mathematical data. Long multiplication for example would have been particularly difficult. Anyone who doubts this should just try multiplying these two numbers without first changing them into their modern forms: MMCDLXIV multiplied by MMMDCCXLVIII.

It was only the later introduction of Arabic numerals that we use today in about AD 1100 that enabled the study and application of mathematics to develop in Europe. This happened as a result of several things. Probably the most important of these was the Muslim conquest of Spain by the Arabs.

Furthermore, Norman knights had in turn conquered the island of Sicily, which had been previously occupied by the Muslims too in an earlier campaign. The knowledge that the Normans had gained from their opponents had also been absorbed by those in Christian Europe who were open to new ideas and eager for information.

There were also a number of crusades when the kings of several European countries had been persuaded by the pope to journey to the Holy Land, along with their armies of knights, in an attempt to free Jerusalem from the Muslim armies that had captured the city. In military terms some of these ventures had been more successful than others; and given the brutality and cruelty that was

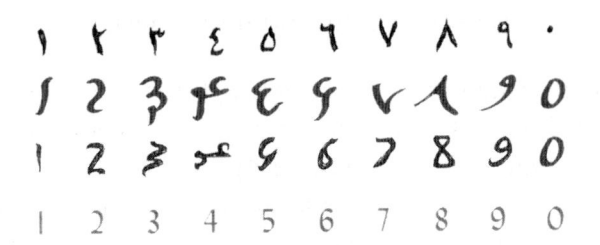

17. Arabic numerals were first used in India. Whereas the Egyptians, Greeks and Romans used different symbols to denote numbers of different value like hundreds, tens and units, Indian mathematicians had devised another method assigning a different value to numerals occupying different columns. This can be demonstrated by comparing how hundreds, tens and units are written in Roman numerals to the modern Arabic forms we use today:

CCCCXXXII = 432; CCXXXI = 231

The Roman numerals have no place value except when used to denote a subtraction as a shorthand way of writing what would otherwise be large numbers: XC = 90 without this device 90 would have to be written with five symbols: LXXXX.

The introduction of the zero to denote an empty column improved medieval mathematics greatly. It was the introduction of this along with Arabic numerals that made it possible for modern mathematics to be developed. This was at first in the form of nothing more than a dot, as seen in the type of Indian-Arabic script above, but this was later replaced by a circle to represent nothing or as we say nought, in later versions. These new numerals were known and used in parts of Europe by some people early in the twelfth century following their introduction from Spain, which was then under Muslim rule, but probably only began to come into general use after 1202 when a medieval Italian mathematician Leonardo Fibonacci published his *Liber abaci*, or *Book of the Abacus*, the first-known European work on Indian and Arabian mathematics.

perpetrated in the name of religion they were hardly justifiable in any terms. Nevertheless, for European society, the gains may be best judged not by the outcome of the conflicts that ensued, but by what was eventually gained as a result of the contact with other cultures, both enemies and allies alike.

These same Muslims that the European armies met on the field of battle were often more advanced in the sciences, medicine, mathematics and astronomy. Despite being at war with one another there was to a degree an exchange of ideas and knowledge from which the Europeans probably benefited the most. Although many manuscripts originating from the remains of the Eastern Roman Empire had already begun to filter through to Europe following the threat of invasion, the Muslim Arabs they were fighting also had copies of manuscripts that originated from ancient Greece itself and the Graeco-Egyptian culture centred on Alexandria, some of which they themselves would have acquired during their conquests of other cultures. They had, moreover, acquired much additional knowledge gained from their own observations and scientific research that were often an improvement upon the more ancient wisdom of past civilisations. It was the contact with the Muslim world and the other cultures the crusaders met on their travels and the arrival of ancient manuscripts that were later brought to Europe by those fleeing the fall of Constantinople, that very likely helped kick-start a new interest in scientific advancement in medieval Europe at the end of the Middle Ages.

The other thing that improved mathematics towards the end of the Middle Ages was the introduction of the zero to denote an empty column. It was this along with Arabic numerals that together, made it possible for modern mathematics to be developed. The reason was that Arabic numerals allowed place value to be given to numbers by putting the units, the 10s, 100s and 1,000s, and so forth, in separate columns. The idea of using a symbol to indicate zero, or nothing, to occupy an empty column was devised by an Indian mathematician. Before, without this symbol, people had to leave just an empty space: for example, if there were no tens between the 100s and the units. Today we would write the number one hundred and six as 106, whereas at one time medieval mathematicians would have had to have been written it without using a zero. This was difficult enough when figures were written in neat columns like this.

1		6
	2	3	4
2	4		5

But without the columns the same numbers, although written in an earlier purely Arabic form, would have often made no sense at all and could have looked much like those shown here: 1 6, 234, 24 5.

Therefore the introduction of Arabic numerals, along with the later introduction of the zero, helped academics of all kinds in their studies and with their observations and recording of natural phenomena. It also made basic arithmetic far easier for ordinary people. By the time of Copernicus Arabic numerals were in common use. If this had not been the case many of the calculations that Copernicus had to do would have been far more difficult and probably more prone to error too. Thus the adoption of these two new systems allowed mathematics and arithmetic to move forward into the modern world.

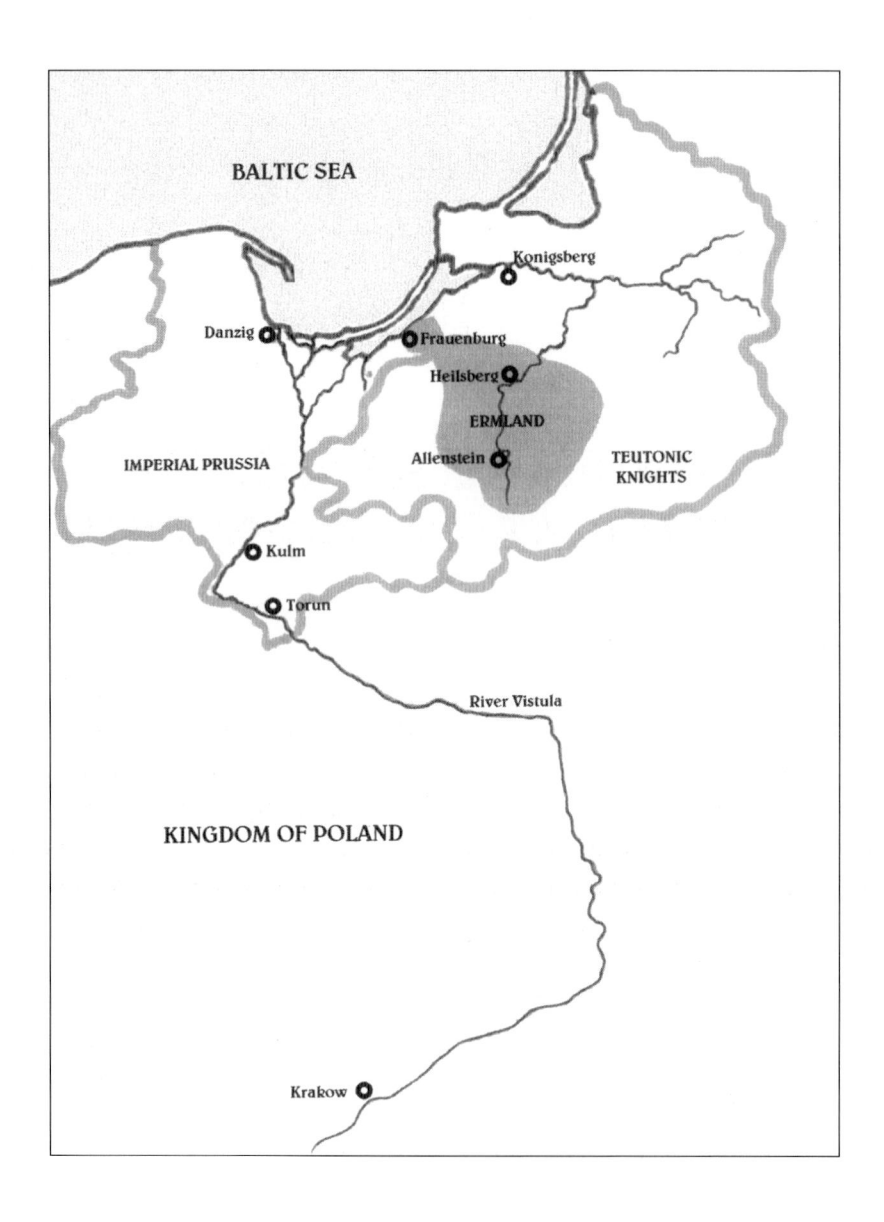

18. Ermland, where Copernicus's uncle Bishop Lucas Waczenrode ruled, was surrounded by the territories controlled by the Order of Teutonic Knights. Recognising the authority of the Polish king as overlord brought the bishop into conflict with the Knights.

IV

THE EDUCATION OF NICHOLAS COPERNICUS

Not everyone who lived in Europe during the medieval period had an opportunity to go to school to be educated. For many children their education was just what they were taught at home by their parents or what they learnt later at work. Some would learn a trade from one of their parents. In the case of girls this often meant how to keep house, cook, sew and bring up a family. Boys might be taught their father's trade, or otherwise if they were lucky, could be taken on as apprentices by a master whose house they would then go to live in while they were being taught a craft or profession. Only the most fortunate though could normally acquire a proper school education. Because the father of Copernicus was a wealthy merchant he was able to send his son to school and the young Nicholas Copernicus began as a pupil at St John's School in his home town of Torun.

But when Nicholas was about ten years old his father died and this could have made it difficult for him to continue his education. Copernicus was though exceptionally lucky. His mother had a brother, Lucas Waczenrode, who at this time was a canon in the Church – and as such held a well-paid and influential position. Nicholas had a brother and two sisters and they would probably have all been destitute if Lucas Waczenrode had not adopted his dead brother's family and taken responsibility for their up-bringing and education. This meant that Nicholas was still able to attend school and obtain the knowledge and basic skills that he would have needed, then as now, to pursue any academic career: which as it turned out he was destined to do.

A few years after the death of Nicholas's father the Church made his uncle, Lucas Waczenrode, a bishop of a region known as Ermland. This was a very important post in medieval times, even more so than it is today, as the Church in those times was much more powerful and influential than now. Indeed Bishop Lucas Waczenrode was also ruler of a part of Ermland and had similar powers as a duke would have had in those days, which were considerable. He would have literally had the power of life and death over those in his diocese. Lucas Waczenrode became guardian of his sister's family. In the process Nicholas Copernicus gained a powerful protector and benefactor whose benign influence he was to benefit from for much of his life.

The bishop was not by any means a soft man and it is said that he was never known to smile; but he had a hard job to do. The duchy of Ermland was one of four vassal states that had been part of the lands conquered by the Teutonic Knights on behalf of one of the Polish rulers, Duke Konrad of Masovia. Even though originally only employed as mercenaries, the Teutonic Knights had not been happy about relinquishing Ermland to Polish rule after the battles they had fought. The finer details of this altercation were described earlier, the matter eventually had to be resolved by armed conflict in which the knights were themselves defeated by the Polish king and his Lithuanian allies. Although by the time Copernicus's uncle became bishop, the Teutonic Knights were meant to give their allegiance to the Polish crown, they were still inclined to dispute Poland's right to rule the lands they had won many generations earlier and there were always storm clouds brewing on the horizon.

His uncle, the bishop, though was good to his sister's family and continued to take responsibility for Nicholas Copernicus's education. We know nothing of his mother after his father death and relatively little about his two sisters, apart from the fact that one of his sisters entered a convent as a nun and the second was married. But when he had finished his schooling Nicholas at the age of eighteen, along with his older brother Andreas, became a student at Kraków University. This university, which was an important and famous place of learning in medieval Europe, had been founded in 1364, more than a century before Copernicus had been born, by Casimir the Great who was the king who had been responsible unifying Poland as one country. Here, at Kraków University, Nicholas Copernicus studied the classic teachings of ancient Greece and Rome.

The rediscovery of manuscripts from these ancient cultures that had been brought to Europe after the fall of Constantinople, along with the knowledge obtained from Muslim sources in Spain and Sicily, had revived interest in the

19. Once his early schooling had been completed, Copernicus at the age of eighteen, along with his older brother Andreas, began his life as a student at the University of Kraków. This probably brought him into contact with some of the best minds in Europe, as Kraków was the capital city of Poland. One of the people the young Nicholas was practically certain to have met as a student was Albert Brudzewski, a famous teacher of mathematics. We do not know if the young Nicolas Copernicus ever attended the lectures of Albert Brudzewski but he must have known of his teachings. It may well have been here that Copernicus first developed an interest in mathematics, which was to later help him so much in his astronomical research.

arts and sciences during the later part of the Middle Ages. Like all the other students in the university he would have had to learn Latin. This was originally the language of ancient Rome and was spoken by all the clerics of the Roman Catholic Church, which adopted Latin as its official language throughout its history. Not only were most of the ancient documents studied at university written either in Latin or Greek but nearly all the books produced in Europe during the Middle Ages were written in Latin too. For this reason every educated person had to learn to read, write and speak Latin if they wished to succeed in an academic career. One advantage of this is that Latin became the common language of all the most learned and cultivated people of medieval Europe.

One of the people the young Nicholas was practically certain to have met as a student was Albert Brudzewski, who was a famous teacher of mathematics. Although there is no definite evidence that Nicholas Copernicus ever attended Brudzewski's lectures he must have known of his teachings. It may have been here, at Kraków, that Nicholas as a young man

20. Medieval schooling: Nicholas Copernicus began his education as a pupil at St John's School in his hometown of Torun but not every child during the medieval period had the opportunity to go to school, as education was costly. However Copernicus's father was a wealthy merchant and thus able to afford a proper education for his sons. Only the most fortunate could normally hope to acquire tuition at a school. For most children their education was just what they could be taught at home by their parents, or what they learnt later at work. In the case of girls this often meant how to keep house, cook, sew and bring up a family. Boys might be taught their father's trade, or otherwise if they were lucky, could be taken on as apprentices by a master whose house they would then go to live in while they were being taught a craft or profession.

first developed an interest in mathematics, an interest that was to later not only help him with his astronomical research but also, more critically, was to enable him to provide the proof required for the theory that was to eventually make him famous.

Other important influences in the young Copernicus's life were the works of Johann Mueller and John Holywood. It is worth mentioning that both these men, like Copernicus and many other academics, had chosen to adopt Latin versions of their names by which they are often better known as this was a very common practise in medieval Europe and if one is unaware of this it can make it difficult to identify some of the people mentioned in other books about the period in which Copernicus lived. For example, Johann Mueller called himself John de Regio or Regiomuntanus, meaning 'King's Mountain' in Latin after his birth place, Königsberg (German for 'King's Mountain'). John Holywood's Latin name was John Sacrobosco, literally meaning sacred wood.

Copernicus most probably read the work of Regiomuntanus when studying astronomy. Regiomuntanus as a scholar had played a major part in collecting together the manuscripts of ancient Greece, including those dealing

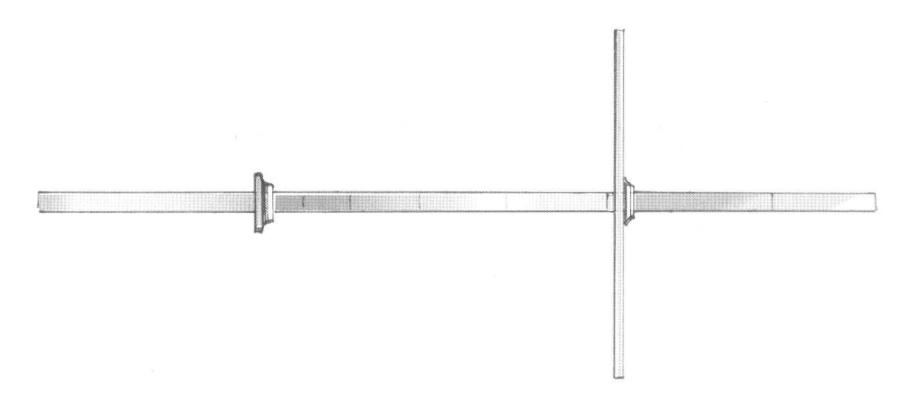

21. The cross-staff was used not only by astronomers but by navigators and surveyors too. Distances and dimensions could both be measured using this instrument, by moving the cross-piece along the staff and taking a reading from the calibrated markings when the correct alignment was made with the object or objects being viewed. The calibrations can be either linear or based upon a trigonometric calculation and with some models the cross-piece can be changed to measure objects of different sizes and a range of distances. It worked on the simple principle that the apparent size of objects and the spaces between them varied with their distance from an observer. Surveyors for example used this to tell how high a hill was and architects the height of a building, whereas navigators could use the same method to determine how high the sun, or a star, was above the horizon to determine their latitude or tell the time of day or night. Astronomers employed the cross-staff in a similar way to track the movement and establish the position of the planets and stars or the sun and moon.

with science and mathematics. Building in part on the knowledge gained from these sources, he wrote one of the first modern books on trigonometry: a form of mathematics that has an important part to play in astronomy. He was also responsible for promoting the use of algebra that could simplify the working of complex calculations. There were other significant contributions too. In 1473, the very year in which Nicholas Copernicus was born, Regiomuntanus had devised a new kind of sundial that could be used to tell the time of day anywhere in the world. This universal sundial was a real innovation as the sun's elevation above the horizon at any particular time is dependent upon several factors. Apart from varying at different times of day its position at any particular time of day changes with the seasons and also at different latitudes, with the sun's apparent height decreasing as you move further from the equator.

One standard textbook the young Copernicus would have read was a work begun by another important academic, the German mathematician Georg Peuerbach. It was a Latin translation from the original Greek book, *Alamgest*, written by an Egyptian astronomer named Ptolemy, who lived in Alexandria about AD 140. Georg Peuerbach himself died before he finished the book and it was completed by Regiomuntanus. Although he died only two years after Copernicus was born Regiomuntanus remained a much respected figure in Copernicus's day. Apart from his own academic achievements he was responsible for collecting together many of the copies of manuscripts from ancient Greece and pre-Christian Rome that had been brought from Constantinople when the city fell to Muslim invaders. It was these same manuscripts that were to largely provide the impetus for the resurgence of learning that we now call the Renaissance.

He was also responsible for a series of very useful astronomical tables known as *Ephemerides,* that showed the daily position of different heavenly bodies, which could be used for navigation. In addition to this he tried to provide a means for sailors to establish their position of longitude by measuring the angle and distance of the moon from various stars. There were, though, difficulties with taking sufficiently precise measurements with the instruments then available and any mariners wanting to use this method of finding longitude would have first needed special training. Providing they knew the time of year, by the beginning of the sixteenth century seamen could calculate their latitude north to south fairly accurately by using a relatively new device known as a cross-staff, or otherwise they used a quadrant, to measure the height of the sun above the horizon at midday. But longitude was different. Before accurate clocks were available at sea there was no reliable way to tell how far east or west one was because the height of the sun at midday, although changing with the seasons, otherwise remained constant at any given latitude however far east or west you travelled – and many ships were lost as a result.

Although John Sacrobosco lived more than a century before Copernicus he too had written a book on astronomy, *The Sphere*, that was still being referred to by students in medieval Europe. It also contained a short section on the theories of Ptolemy and some pertinent observations about reckoning time and comments on the shortcomings of the Julian calendar. A further section, entitled *Algorismus*, is particularly note-worthy because it also tells the reader how to employ the then newly introduced Arabic numerals to carry out their astronomical calculations.

The translations by Georg Peuerbach and Regiomuntanus, as with John Sacrobosco's book, may have been Copernicus's earliest contacts with the Ptolemy's model of the cosmos. Ptolemy was to figure significantly in the life of Copernicus. The wisdom of ancient civilisations was held in high regard in Europe during the Middle Ages and because of this Ptolemy's ideas were accepted without question by most people at that time. Ptolemy believed that earth was at the centre of the universe, because the sun, moon and stars and all

22. The wisdom of ancient civilisations was held in high regard in Europe during the Middle Ages and because of this the ideas of astronomers such as Ptolemy of Alexandria were accepted without question by most people at that time. Ptolemy believed that our planet Earth was at the centre of the universe, because the sun, moon and stars and all the other planets seem to revolve around us. But Copernicus had growing doubts about the Ptolemaic geocentric planetary model when an experiment he helped conduct as a student suggested that Ptolemy's theory regarding the movements of the moon was based upon a false premise.

the other planets seem to revolve around us. But it was these very ideas that Copernicus as an astronomer was eventually to challenge and attempt to disprove. This was to make Nicholas Copernicus himself famous and is also why he is remembered to this day.

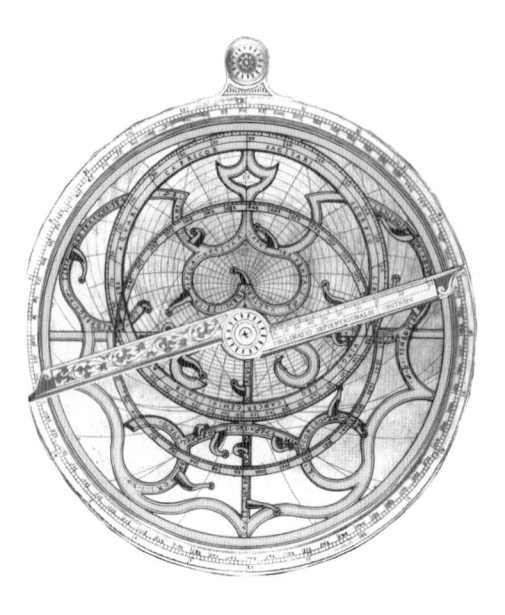

23. An astrolabe is basically a set of disks with either one or two moveable double-ended arms that can be adjusted to measure, among other things, the altitude of heavenly bodies such as the moon, stars or planets. They were sometimes employed by seamen as an aid to navigation and by mapmakers and surveyors to measure distances, elevations and angles as well as by astronomers.

This device consists of a circular backplate or *mater* displaying a calendar around its rim with the signs of the zodiac and showing both the months of the year and days. A second plate lying on the backplate was engraved with curved lines of longitude and latitude and circular lines showing the Equator and the Tropics representing the spherical surface of the earth. On top of this was a moveable fretwork disk called a *rete* that had pointers representing the most prominent stars. This was turned until the star pointers lined up with the appropriate settings corresponding to the correct date before taking a reading. The double-ended arm behind the backplate could be rotated to line up (using sights) with the object being viewed. A second double-ended arm above the *rete* was used to obtain additional information such as the latitude (or declination) of various stars or to calculate the time (reproduced with the kind permission of the National Maritime Museum, Greenwich).

V

ASTRONOMICAL INSTRUMENTS

It was probably while at Kraków University that the young Copernicus first developed an interest in astronomy. During his final year four scientific instruments were delivered to the university. They came from Buda in Hungary and were a gift from a well-known Polish astronomer Marcia Bylica who had once been a student at the university. It is said that the rector of the university was so excited when they arrived that he arranged a special assembly for the students to show them the university's new acquisitions. They were a celestial globe, two astrolabes and a triquetrum. These instruments, which some readers have probably never heard of, would seem in some ways very basic by today's standards but at the time they represented the very latest in astronomical equipment.

An astrolabe is basically a set of disks with either one or two moveable double-ended arms that can be adjusted to measure, among other things, the altitude of heavenly bodies such as the moon, stars or planets. Although appearing to be a more simple device in its basic construction, it was nevertheless a very complex and refined instrument inscribed with measurements that had to be produced with great accuracy. Some of these measurements were marked on a circular backplate known as a *mater* and included a calendar with the signs of the zodiac and showing both the months of the year and divisions for each day. There was also a second, upper plate that was located on the *mater*. This was engraved with lines of longitude and latitude that were curved as the lines on some maps are when representing the

24. An astrolabe can be used for several purposes and was employed by architects, surveyors and map-makers, as well as travellers who wanted to establish their location or direction, such as seamen, and was also employed by astronomers. Normally the astrolabe was held in the way shown on the right and both sites were then lined up on the object being viewed, but when taking a reading from the sun it was held by one's side to avoid damage to the eyes. The bar behind the backplate of the astrolabe, called the *alidade* was then rotated until the sun's rays passing through the first site fell on the second lower site. This enabled the observer to measure the elevation of the sun without risking their sight.

spherical earth on a flat surface. Circular lines showing the Equator and the Tropics of Cancer and Capricorn were also marked on the upper plate. On top of this was a moveable fretwork disk that had pointers representing the most prominent and easily identifiable stars. This upper disk, called a *rete*, could be turned to line up the star pointers with the appropriate settings corresponding to the correct month and day in the year before taking a reading.

Behind the back-plate there was a double-ended arm, or bar, called an *alidade*. This revolved and to take a reading the astrolabe was held upright, edge on the object being observed and readings were taken by adjusting the angle of the double arm until the sighting vanes set at each end of the *alidade* could be lined up with the object being viewed. If this was a star or planet this was done by looking through the sights; but when it was the sun a visual sighting could cause damage to the eyes and even result in the observer being blinded (many centuries later, the regular use of a sextant by young navy midshipmen, to find their ship's position by measuring the altitude of the midday sun, resulted in many of them being left blind in one eye long before they ever became lieutenants). Instead the astrolabe was held at waist height and then the *alidade* was simply rotated until the sunlight passing through the upper sight fell directly on to the centre of the lower sighting vane. Then there was usually a second, revolving, double-ended arm above the upper plate known as a *rule* that was used to obtain additional information from the *rete* such as the latitude, or declination of various stars.

Each individual upper plate was only designed to be used at a particular latitude. Each time one travelled north or south the astrolabe had to be dismantled and a new plate inserted on top of the *mater* with different measurements before it could be used. This was a very simple operation and astrolabes were often supplied with several plates for this purpose. The main uses of an astrolabe in these times was to take readings that could used in astrology and astronomy. It could be used to tell the time too by reference to the stars, and from the sun too if you knew the date, and was sometimes employed by seamen as an aid to navigation. Because it could measure angles very accurately it was also often used by mapmakers and surveyors. Indeed it was quite usual for astrolabes to have additional scales and measurements on the back of the *mater* to help with surveying.

The second kind of instrument, a triquetrum, could be used in a similar way to an astrolabe. It was far less portable and did not have the same range of scales and measurements found on an astrolabe; but it could also be employed to measure the angle between different celestial objects, as well as giving their height in degrees above the horizon. This was the same kind of instrument had been used centuries before by Ptolemy and the construction of the triquetrum was described by him in his manuscripts which were still being referred to by astronomers during the Middle Ages.

25. The armillary sphere is an instrument that models the celestial sphere with respect to the horizon of an observer unlike the celestial globe mentioned in the text. It is made up of rings (known in Latin as *armillae*) representing the great circles on the celestial sphere such as the horizon, the celestial equator, the tropics and the ecliptic. They can be adjusted to any latitude and can be used either as an instrument of observation or as a tool of demonstration. The earliest known description of such an observational armillary sphere was given around AD 150 by Claudius Ptolemy in his *Almagest*. Ptolemy's instrument, which is known as a zodiacal armillary sphere, measured co-ordinates with respect to the ecliptic (zodiac); i.e. the rings were calibrated to read latitudes and longitudes in relationship to the ecliptic, that represents the apparent path of the sun.

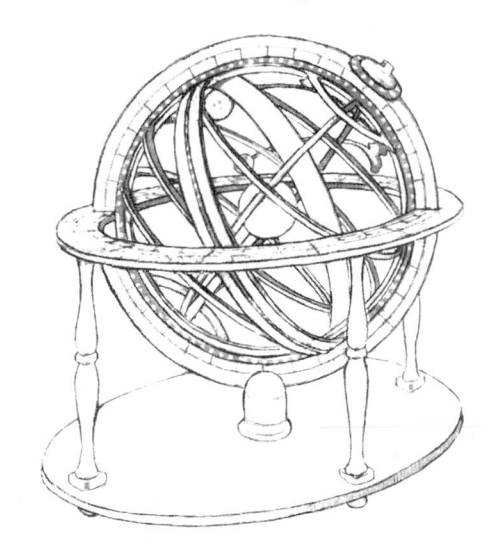

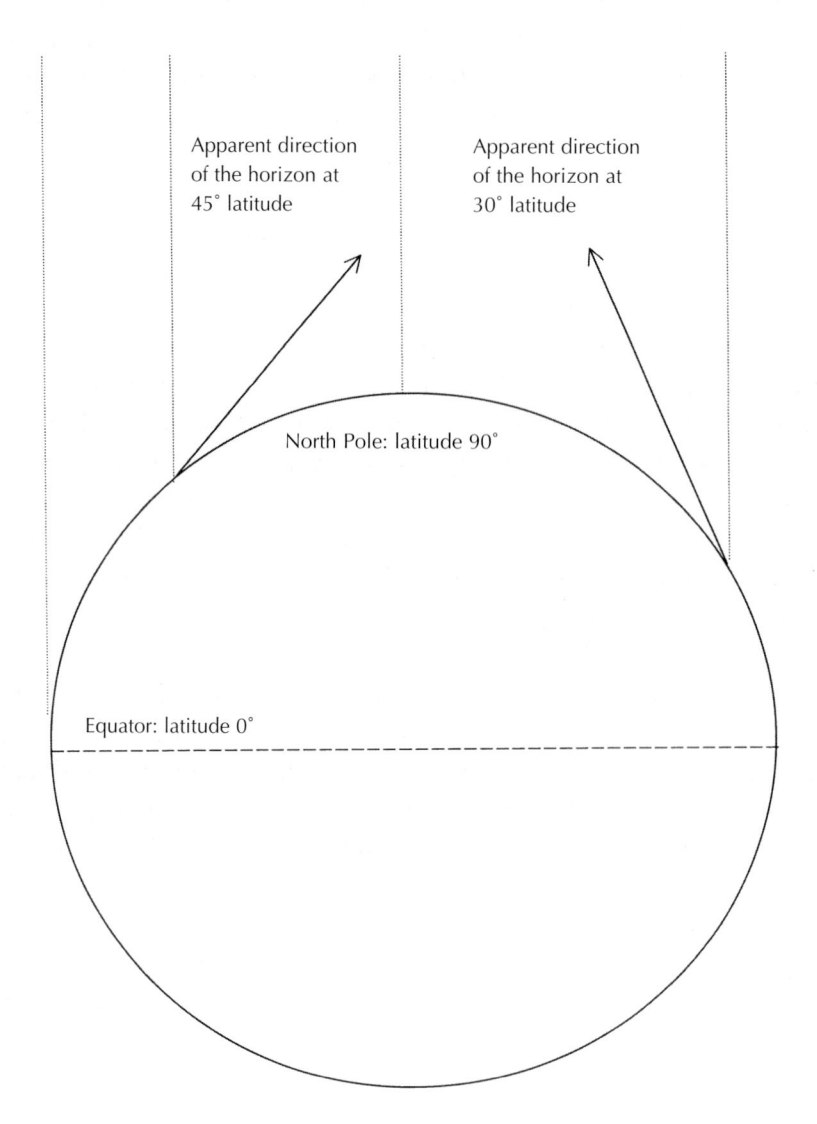

The perpendicular lines below show the direction in which the Pole Star, *Polaris*, would appear at different latitudes

Apparent direction of the horizon at 45° latitude

Apparent direction of the horizon at 30° latitude

North Pole: latitude 90°

Equator: latitude 0°

26. The perpendicular lines each show the direction in which the Pole Star will appear in relationship to the horizon at differing latitudes. Standing at the North Pole the Pole Star, *Polaris*, would be directly overhead and therefore at a right angle, 90 degrees to the horizon, whereas at the Equator the Pole Star would be on the horizon and the reading would be literally 0 degrees. This meant that at night navigators could determine their latitude by measuring the height of the Pole Star above the horizon. For example at 45 degrees of latitude *Polaris* would appear 45 degrees above the horizon and at 30 degrees of latitude the apparent altitude of *Polaris* would be at 30 degrees. During the time of Copernicus seamen would have used either a quadrant or astrolabe to measure the altitude of *Polaris*. Astronomers employed versions of the same instruments to determine the position of stars and other celestial bodies. Later in the sixteenth century another newer, simpler instrument called a cross-staff was introduced which was also employed in much the same fashion.

This instrument consisted of three straight bars with markings which were used to take the measurements and for this reason the triquetrum was sometimes known by another name – Ptolemy's rulers. The celestial globe was similar to the normal terrestrial globe which we are still familiar with today, that shows the ocean and continents of the earth; but instead a celestial globe is marked with the positions of the more noticeable, visible stars displayed in the constellations they form. The globe itself would have been set at an angle just as with a terrestrial globe. This represents the angle of axis about which the earth rotates in relation to the sun and there would also have been a fairly broad horizontal band encircling the globe, most probably showing the months of the year and signs of the zodiac. Many terrestrial globes at the time would have had the same thing as they were often produced together as a pair. This marked what is called the ecliptic that actually represents the plane of the earth's orbit around the sun, but during Copernicus's time, when most people believed the earth was the centre of the universe, it was thought to be the angle at which the sun orbited the earth. It was this very misconception that Copernicus was to later challenge and eventually disprove.

There were several other astronomical instruments in use during the Middle Ages apart from the ones presented to Kraków University. The most simple of these to use was probably the cross-staff, first mentioned in 1342, its use gradually became commonplace from the beginning of the sixteenth century. This device often consisted of no more than a wooden cross with a sliding crosspiece that could be moved up and down the upright. Both the apparent distance between celestial objects could be measured using this, or their height above the horizon: to assess the position of the sun to obtain the time of day. Among other things it enabled mariners and others to discover their latitude.

This involved nothing more than measuring the apparent distance between the horizon and the Pole Star. The observer pointed the upright of the cross

27. Equatorial sundials can be adjusted to give the time at different latitudes. In the Northern Hemisphere model here, the ring showing the times is raised or lowered against the marked scale to set it to the appropriate latitude, then orientated to face north by using the built-in compass. The small plumb at the back is first consulted to ensure that the device is upright before taking a reading. All the elements mentioned and the *gnomon* can be folded down flat to allow the sundial to be easily transported. Small enough to fit in the palm of the hand, it was the medieval and Renaissance equivalent of the pocket watch.

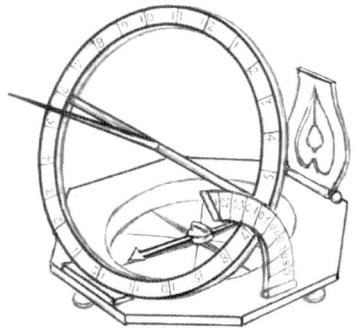

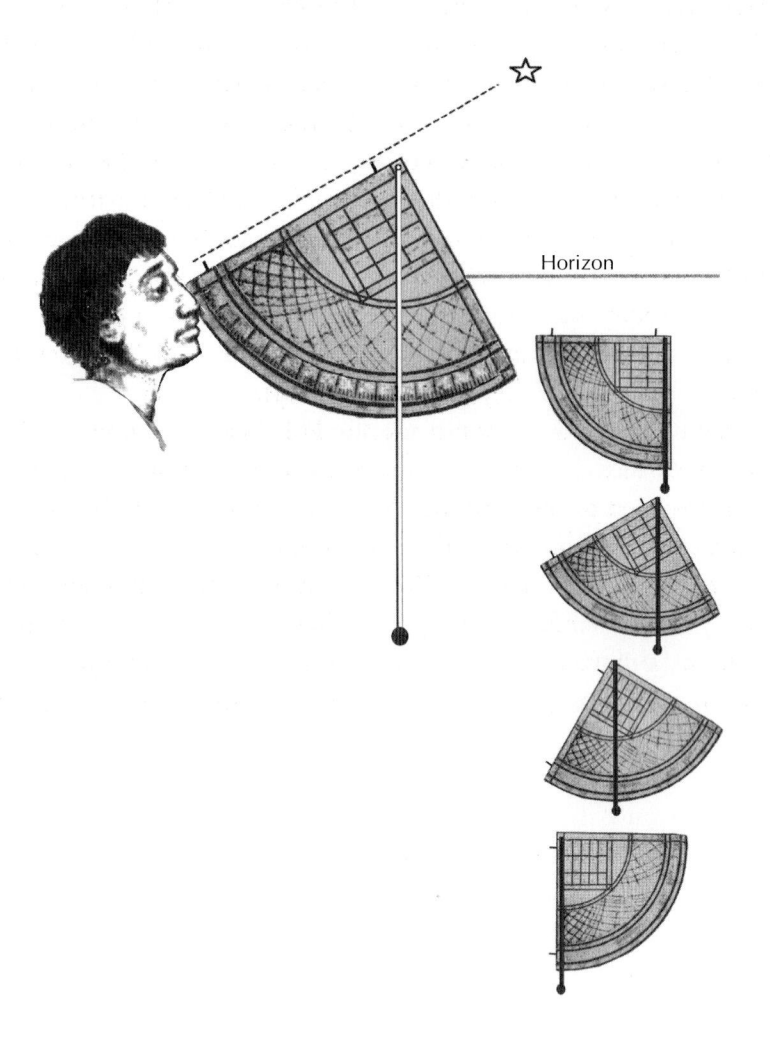

28. When using a quadrant the observer puts the device against his or her cheekbone and uses the sights to line up the quadrant with the object being viewed. They then have to note the position of the plumb line, against the scale inscribed along the curved edge of the instrument, to discover the elevation of the target object above the horizon: or in other words, its degree of arc from the horizontal. The four quadrants on the right demonstrate the way in which this works. The first is level with the horizon, the second is at 30 degrees from the horizontal and the third is angled at 60 degrees. The fourth quadrant shows the position it would be employed in to view an object directly overhead, at 90 degrees. Navigators used the quadrant in this way to establish their position and architects, surveyors, as well as astronomers to measure angles of elevation.

towards the northern horizon and then slid the crosspiece along the other member until one end lined up with the Pole Star and the other with the horizon. It was known that the apparent distance between the two would decrease as one travelled south and therefore, the more the distance decreased the further the crosspiece had to be moved away from the eye of the observer, before the two ends would line up with the Pole Star and the horizon. This means of measurement relied entirely upon the principles of perspective, which ensured that the crosspiece would also appear to become smaller as its distance from the observer increased. The latitude could then be determined by taking a reading from the position of the crosspiece. The further the crosspiece was from the eye the further south one was.

Once a ship had travelled too far south and the Pole Star had dropped below the horizon, another method had to be employed using an instrument called a quadrant. This was used to measure the height of the sun at midday, which, dependent on the time of year, would also vary as one moved further north or south. In fact due to the tilt of the earth, the elevation of the sun would only be at its highest point towards the equator during the spring and autumnal equinoxes. Both these devices, as with the astrolabe, were employed for a number of other purposes by people like surveyors and map-makers to judge distances and elevations.

29. Bishop Lucas Waczenrode as ruler of Ermland would have literally had the power of life and death over those in his diocese. Upon the death of Copernicus's father Lucas Waczenrode became guardian of his sister's family. In the process Nicholas Copernicus gained a powerful protector and benefactor whose benign influence he was to benefit from for much of his life. The bishop took responsibility for the education of Nicholas and his brother Andreas and found them both positions as canons of the Church. Nicholas Copernicus was also later to act as his uncle's physician and private secretary.

VI

THE BISHOP'S NEPHEW

His uncle Lucas, as such a powerful man, was able to help Copernicus a great deal. The first thing he did, once Nicholas had completed his studies at the university in Kraków, was to begin making arrangements for him to become a canon of the cathedral at Frauenberg. This may seem a strange choice to make for a younger man. A canon was a kind of monk and this form of life would not have been to everyone's liking as canons like priests were not allowed to marry. We do not know what Nicholas thought of it, but it is doubtful that he could have gone against his uncle's wishes if he had wanted to, as the head of the family in those days was practically always obeyed. His uncle was not being unkind either, as he knew that there would be several very important advantages that his nephew would gain from this. And he was to be proved correct, as some of these advantages Nicholas Copernicus was to benefit from for the rest of his life.

The most significant advantage was that as a canon of the Church he would have a job and an income for life. This was to give Nicholas a financial independence that few other people would have enjoyed then. It would also form a basis for his further education because most of the libraries of medieval Europe belonged either to the universities or to the Church – as did the vast majority of books. This especially applied to the great classics of ancient Greece and Rome as well as the writings of the Arab world. Furthermore, it was within the Church that he would meet many of the great thinkers and philosophers of his own day, because this

Estimating the distance of celestial objects

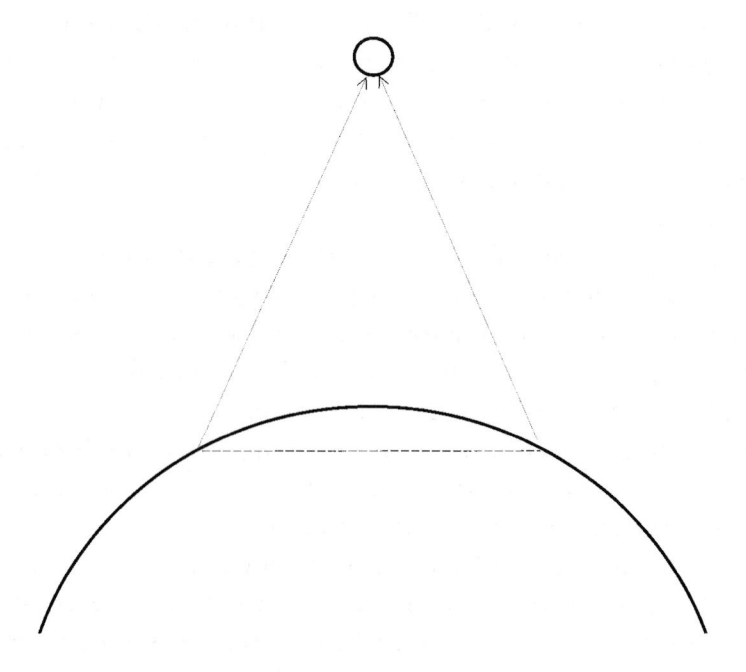

30. Parallax is the apparent change in the position of an object when it is viewed from different positions. It is sometimes referred to as trigonometric parallax because it relies upon the known properties of triangles to assess distances. Since ancient times astronomers have tried using this method to measure the distance of such celestial bodies as the sun or the moon and the other known planets. The true extent of the solar system and the universe were never truly comprehended by the astronomers of the ancient world, who did not realise that most of these celestial bodies were too far away to obtain useful results using the methods and equipment they had available. When calculating the parallax of a celestial body vast distances are involved, which means that any error however small would be greatly enhanced.

was still one of the few institutions in which the majority of members were both literate, numerate and, moreover, well educated by the standards of those times.

Frauenburg Cathedral was the official seat of Bishop Waczenrode and so it was unlikely that there would have been much difficulty in finding Nicholas a post there as a canon. But there was no immediate position vacant and so Nicholas was sent by his uncle to the University of Bologna to continue his studies. He was later joined by his older brother Andreas who was also enrolled as a student there.

One of the subjects Nicholas studied at Bologna University was canon law, which involved learning about the legal obligations and powers of the Catholic Church, which in those days were considerable. This was to prepare him for his career as a canon at the cathedral. The main tasks of a canon were to help the bishop both in the day-to-day running of the cathedral and in the administration of the diocese. It was therefore important that Nicholas knew the law well and had a good understanding of all the legal issues and implications that may arise while carrying out his duties.

Canon law was not though Copernicus's main interest, or the only focus of his studies. It was the mysteries of astronomy that most intrigued the young man and he was fortunate enough to find the right teacher to help him pursue this interest. Professor Domenico Maria da Novara was not only a famous astrologer; he was also a compiler of astronomical tables and the time Copernicus spent at Bologna University probably helped formulate many of his own theories about the nature of the universe.

It is thought that Copernicus learnt Greek at this time. It would have meant that he could decipher ancient documents for himself and discover the real meanings of what they contained by reading them in the original language. He would have discovered that the followers of Pythagoras for example did not believe that the earth was the centre of the cosmos around which all other heavenly bodies revolved, but thought that it was in motion like all the other planets. He would have known too that da Novara had doubts about Ptolemy's geocentric model of the cosmos.

His teacher also became the young Copernicus's friend and even took him in as a lodger in his own house. The two men, student and teacher, collaborated in their astronomical research and Copernicus must have learnt a great deal during this period while working with such an experienced mentor. The professor and his student measured the altitude of stars and made observations together to assess the distance of the moon. They did this by calculating what is known as the moon's parallax. This involved taking observations of both the full moon and the quarter moon. By comparing the information gained during these two different phrases it was then theoretically possible to work out how far away the moon was from the earth's surface. More importantly the experiments he did with Professor Domenico da Novara, helped sow in Copernicus's mind growing doubts about the Ptolemaic geocentric planetary model as discrepancies in Ptolemy's predictions suggested that his theory regarding the movements of the moon and the other planets may have been based upon a false premise. The date was 9 March 1497 and discovering evidence of Ptolemy's inadequacy was a pivotal moment in the life of Copernicus.

Much of what he learnt while a student at Bologna University was to be used by Copernicus in his life as an astronomer and would also help later to support his own ideas about the cosmos. Even so it was not all work during his student days at Bologna as it is known that Nicholas and his brother Andreas, who both had an income as canons of the Church as well as allowances from their uncle the bishop, had nevertheless managed to accumulate some fairly serious debts before their final year had finished and had to borrow money to pay them off.

When the year 1500 came the Church chose to mark the turn of the century with a jubilee in Rome. Christians everywhere were invited to attend the celebrations and monks, nuns and priests as well as many ordinary people flocked to the city. Among them were Nicholas and his brother Andreas, who may have had an official role to play as representatives of the Ermland chapter where their uncle Lucas Waczenrode was bishop. It is said that more than 200,000 others were there that year in St Peter's Square to hear the pope's traditional Easter Sunday blessing.

The city of Rome was the focal point for the whole of Christian Europe, where people turned for both leadership and religious guidance. But as with all centres of power there was corruption, greed and unbridled ambition. Popes were not immune to temptation either. At the time they wielded immense power and were often inclined to make up their own rules of conduct. For example they were not by any means always celibate as today. The incumbent, Pope Alexander, had a family. As both a cardinal and pope, Rodrigo Borgia (he took the name Alexander on becoming pope) fathered a number of children by his mistress Vanozza Catanei. Scheming and intrigue were rife among the family members and while the centennial festivities continued the pope's own son, the notorious Cesare Borgia, was to have the husband of his sister Lucrezia murdered. He was never called to account for this crime. In fact as Duke of the Romagna he went from strength to strength. Being made

31. According to a friend of his later years, George Rheticus, the young Nicholas Copernicus gave a lecture on mathematics while visiting Rome in 1500 for the jubilee celebrations (adapted from *Vie des Savants Illustres,* Figuier, Louis, Hachette 1883).

captain-general of the armies of the Church he was able to help bolster his father's papacy and later even attempted to establish his own principality in Italy. He was very probably a major influence upon the writer Machiavelli, as a model in his notorious and contentious book, *The Prince*, that describes how one can best both gain power and retain it as a head of state.

At that time Rome appears to have been a particularly dangerous place to visit or live. According to one contemporary account there were on average four or five murders every night and not only of ordinary folk: we are told that both bishops and prelates were also among the victims. There were also many public executions with eighteen people on one occasion being put to death in a single day when they were hung together from the gallows on the Angel Bridge. The miscreants included a physician and surgeon from the local hospital who had taken to supplementing their income by robbery and murder.

Fortunately Copernicus seems to have avoided being embroiled in any of the dangerous intrigues that were obviously going on while he remained in Rome. He stayed there for about a year and in that time appears to have given several lectures on mathematics and astronomy to students. He also took the opportunity to observe an eclipse of the moon on 6 November 1500. Then, he and Andreas departed, having an important appointment to keep at Frauenburg where the two brothers were to be officially installed as canons of the cathedral.

32. Despite his pious appearance, Pope Alexander VI, the erstwhile Cardinal Rodrigo Borgia, was probably the most notorious of all the popes. Not only did he keep a mistress, Vanozza Catanei, and fathered several children by her, he also used his high office to bestow favours on his offspring and protect them – even from prosecution for murder. His daughter Lucrezia Borgia was a suspected poisoner and his son, Cesare Borgia, quite apart from his excessive political ambitions, was a known murderer. In the year 1500 during the celebrations organised by the Church to mark the turn of the century, he had his brother-in-law, the husband of Lucrezia, killed. His father ensured that he was never called to account for his crime.

After their inauguration as canons at Frauenburg Cathedral both the brothers were given permission to continue their education in Italy before taking up their appointments. Andreas returned to continue his studies in Rome, but when Nicholas left Frauenburg it was not to return to Bologna University but to attend the university at Padua. This sort of change of venue was not at all unusual in the Middle Ages as students were often encouraged to swap between different universities during their studies. This was probably because it was thought to provide them with a broader and more varied education. Furthermore, it was at Padua that he was later destined to study medicine and this may have had a bearing on his decision to change universities. Whatever the reason, in 1501 Copernicus transferred to Padua University where he continued his studies in law.

Even then, when it was time to take his final examination Copernicus did so not at Padua, where he had completed his studies, but made a further move to another university in Ferrara. There may have been a sound reason for this, as the rituals associated with being granted a doctorate involved lavish celebrations to which all one's friends and fellow students had to be invited. This could be very expensive and as we know Copernicus and his brother had already had one experience of running up large debts. It may not have been coincidental therefore that students quite frequently chose to take their doctorates at a different university to the one at which they had studied, where

33. Machiavelli served both as a secretary and diplomat to various dignitaries, the most famous being the son of Pope Alexander VI, the notorious Cesare Borgia. He was also for a time in the service of the Medici. When he eventually fell from grace and retired he became a writer. He is in fact best known as the author of *The Prince*, which describes how one can best obtain and maintain power as a head of state. It is said that he largely based it upon the ruthless and self-seeking exploits of his erstwhile employer, Cesare Borgia. Machiavelli himself, although ambitious, was not as wicked as his book might suggest. When Cesare Borgia was eventually overthrown upon the death of his father and protector, the pope, and imprisoned for his crimes, Machiavelli thought that it was a punishment richly deserved.

there would be few if any friends and acquaintances to attend the final ceremonies and merry-making. All that is known is that in 1503 Copernicus was awarded his doctorate of canon law at Ferrara University.

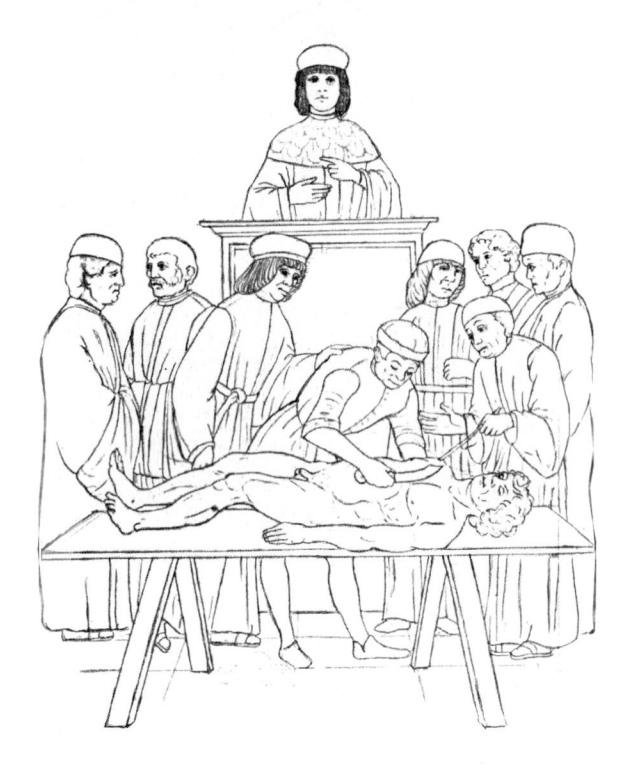

34. This is a typical medieval dissection which would normally be carried out on bodies of executed criminals and used as a teaching aid when training medical students. The doctor usually sat on a rostrum overlooking the proceedings and gave a lecture to the watching students about the anatomy of the various parts of the human body revealed as the dissection progressed.

VII

MEDICINE AT PADUA

In many respects it would seem that Copernicus was destined to continue as a student for a large part of his early life. Once Copernicus had finished his legal studies he returned to the University at Padua and began studying medicine. This was not because he had any intention of being a doctor and there is no record of him taking a medical degree, but it was thought a good idea for members of the Church to have some knowledge of medicine so that they could help the sick of any community in which they served, as well as administering to their spiritual needs.

Copernicus had in fact asked for permission to continue his education at university before taking up his duties as a canon at Frauenburg Cathedral and this request had been granted on the conditions that he included medicine in his studies. Much of medicine in those days was basically book-learning and included very little practical experience of the kind trainee medical students are given today. Nevertheless, what little evidence there is suggests that Copernicus was well respected as a medical practitioner.

There were some aspects of Copernicus's medical training that we would still find familiar today. One of these was the practice of dissection that was carried out on dead bodies. In medieval Italy this would have normally been on the bodies of executed criminals and like a number of dissections done nowadays it was used as a teaching aid when training medical students. The dissections were not carried out by a doctor, however, as they are today. Doctors in medieval Europe were considered very important people and few

35. This illustration is based on a carving from the University at Bologna, in Italy. Captured in stone nearly four centuries ago, these students seem little different from the students of today. Universities were first established in Europe during the medieval period, tended to specialise in particular subjects. Bologna University was renowned for medicine and legal studies. It was one of several universities that Copernicus attended as a young man, when he was sent by his uncle Bishop Lucas Waczenrode, to acquire a degree in canon law.

would have dirtied their hands by doing anything so distasteful as cutting up a body. This was done by special assistants under the direction of the doctor who normally sat aloof on a rostrum over looking the proceedings and gave a lecture to the watching students about the anatomy of the various parts of the human body revealed as the dissection progressed.

Not all the students were men. There were a few women doctors even in the Middle Ages. One female student, Alesandra Galioni, is credited with being the first person to have injected coloured dyes into the arteries of cadavers to trace the path of the blood vessels. She was a student of Mundinus of Bologne who was one of the most famous anatomists of his day. Mundinus is also said to have been one of the first physicians to have actually descended from his rostrum to join the students during dissections. He did so in order to check the anatomical observations of Galen who had been a doctor in Rome during the second century.

Mundinus himself was responsible for writing an anatomical textbook, based on the regular dissections he supervised. Written in 1316, the first printed edition was published in 1478 at Padua, the university attended by Copernicus. Nevertheless, many of the lectures there would have still been based on the writings of the second-century Roman doctor Galen. These would have formed a very important part of Copernicus's medical studies as Galen's work remained a standard text for all medical students throughout the Middle Ages. The manuscripts written by Galen were also illustrated with drawings that attempted to explain the anatomy of the human body and these were often referred to by medieval physicians. Unfortunately Galen did not understand the function of the organs as well as we do today so his manuscripts were sometimes quite misleading and could have led to serious mistakes being made and on occasions poor if not downright dangerous treatment being given.

There is no doubt that Galen was a genius in his day and an excellent observer who wrote 300 treatises on medicine, many of which survived, but no-one had been allowed to dissect human corpses during the times in which he lived and most of his ideas about human anatomy were based on his dissections of animals. This often led to errors and misunderstandings. His information about anatomy was gained from the bodies of such animals as dogs, cows and pigs that often have a different arrangement of organs to humans and in some cases very different digestive systems. The nearest animal to a human that he dissected was a Barbary ape, which is a type of African monkey.

In fact patients frequently died in the Middle Ages due to the limitations of medical knowledge and the incomplete comprehension that doctors had of how the body really works. The true function of the heart was not understood, nor the lungs and other major organs, even though Galen's writings make it obvious that he had made a careful and very thorough study of these organs in the animals he had dissected.

During Copernicus's time at Padua, one of his teachers was Marcus Antonius de la Torre who had commissioned anatomical drawings of both men and horses from the famous artist Leonardo da Vinci. Fortunately there are still many examples of Leonardo da Vinci's work in existence today. Queen Elizabeth II is guardian of several sketchbooks. We can assume that these drawings would have been far more accurate than those derived from Galen as it is known that Leonardo's anatomical drawings were done from life – or more accurately dissected cadavers.

Another major source of information that influenced the practice of medicine was the teachings of Hippocrates, a Greek physician who taught and practiced medicine on the island of Kos, his place of birth, as well as elsewhere in Greece, including Athens. In fact, although he was born 460 years before Christ, doctors worldwide, even today, are still expected to abide by the same oath that Hippocrates made his students swear nearly 2,500 years ago.

But medieval medicine did not rely entirely upon the reiteration of ancient wisdom. Original work was undertaken and new ideas about the treatment of disease and the care of patients were proposed. More than 100 years before Copernicus became a student, back in the thirteenth century, William of Saliceto (*Guglielmo de Saliceto*) had given lectures on health and hygiene that

36. Although medieval physicians had a limited understanding of how the human body worked they were not without skill, nor entirely lacking in knowledge. Quite complicated surgical procedures were successfully undertaken at times and the setting of broken limbs and dislocated joints were just some of the tasks effectively undertaken in the course of their work.

were still well in advance of the times even in Copernicus's day. Among other things he advised women to take particular care maintaining cleanliness during pregnancy and when looking after very young children: recommending that they should be bathed each day. None of these things seem particularly novel to us nowadays but then such ideas were almost unknown. He also advised against allowing children of any age to drink wine and understood the importance of good nutrition and regular exercise. He seems to have realised too that some diseases were contagious and gave good advice about prevention. Unfortunately these ideas were not readily adopted and many otherwise avoidable contagions in medieval Europe still went largely unchecked as a result.

Theodoric Borgognoni of Cervia was another equally influential figure at the University of Bologna. Although not a contemporary of Copernicus he was one of the most respected surgeons of his day, he may have even anticipated Lister in the precautions he took to avoid septicaemia. Not only did he insist upon basic cleanliness; he also used wine to wash the wounds of his patients and lint soaked in wine to hold the edges of wounds together while they healed. Continuing a technique employed by his own father, another great surgeon, known as Hugh of Lucca, he used sponges soaked in opium and hemlock, hyoscyamus, lettuce and mulberry juice, which could be applied to the operational sites to relieve pain and to the nose of the patient to induce drowsiness during surgery. Despite such soporifics being known about they were not widely utilised and surgery remained in most cases excruciatingly painful for the patient throughout the Middle Ages and something to be feared and avoided at all costs.

There were a number of complaints such as haemorrhoids, fistula and scrofula, that is actually tuberculosis of the lymph glands in the neck, which could be successfully treated by surgery in the Middle Ages as well as the removal of cataracts, providing infection could avoided. Without effective anaesthesia the pain at times must have been unbearable. Various constraints had to be devised for each type of operation to prevent the patient moving, if the surgeon was to avoid causing them additional injury. Nevertheless the amputation of limbs infected by gangrene was common-place and even operations to treat breast cancer were attempted, often with good results. Even the most agonising surgical procedures were quite often carried out with the patient fully conscious and without anything to deaden the pain.

One reason may have been that although various soporific and other concoctions to relieve pain or to reduce consciousness during surgery existed they were sometimes in themselves very dangerous. One of those described contained lettuce, gall from a castrated boar, bryony, opium, henbane, and

3 times natural size
Hemlock, *Conium*

Henbane,
Hyosyamus niger maculatum

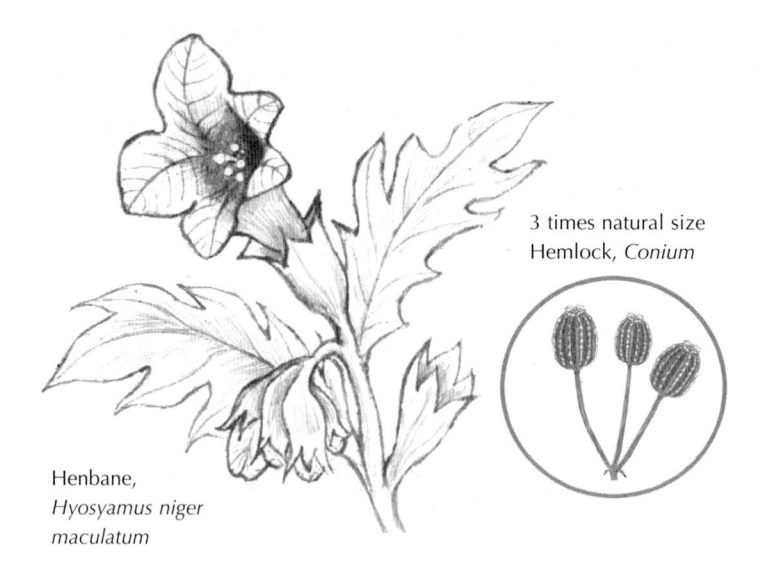

37. Hugh of Lucca and his son, Theodoric Borgognoni of Cervia, as surgeons both used sponges soaked in opium and hemlock, henbane (*Hyoscyamus spp.*), lettuce and mulberry juice, that they applied to operational sites to relieve pain and to the nose of the patient to induce drowsiness during surgery. Another mixture described contained lettuce, gall from a castrated boar, bryony, henbane and opium. They were, effectively, the medieval equivalent of modern anaesthetics.

hemlock juice. Both the bryony and the hemlock juice were potential lethal poisons and great care would have had to have been taken in their administration to avoid killing the patient. Even worse some recipes included such unpleasant ingredients as ground earthworms, urine, and even animal excrement. Medieval remedies often employed hundreds of different substances in the belief that every substance in nature could exert some form of influence.

Although most of the medical men mentioned here lived long before Copernicus became a medical student, many of his contemporaries would have still considered such ideas extremely revolutionary. There was undoubtedly great ignorance about the causes of illnesses and their treatments during the Middle Ages, and without antibiotics many diseases simply could not be treated effectively if at all by even the best physicians. But equally it has to be recognised that there is also compelling evidence for some good diagnostic skills and occasional, isolated examples of sound practice especially among the later physicians of that period.

Both the physicians referred to had been professors during the thirteenth century at Bologna University; the medical school at Padua had been largely modelled upon that at Bologna and owed its existence in part to the constant migration of students that occurred between universities during the Middle Ages. There is little doubt that by the time Copernicus was beginning his studies, the University of Padua was already held in equally high regard as that of Bologna and would have provided Copernicus with one of the best medical educations available in medieval Europe.

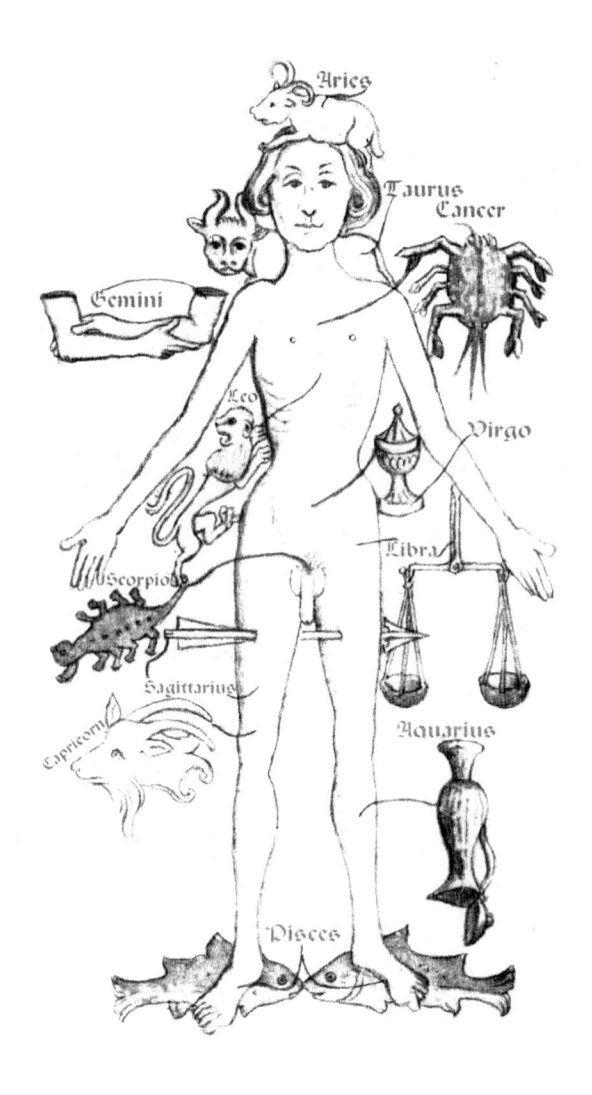

38. During the Middle Ages medicine was closely associated with astrology, which at that time was barely distinguishable from the science of astronomy. It was the astrologer's belief that the sun, moon, planets and stars also affected a person's health and different parts of the body were also associated with different signs of the zodiac, as indicated by the drawing above. Moreover, it was thought that the kind of ailments people were prone to was influenced by time of the year they were born and the position of various stars, planets and both the sun and moon at their actual moment of birth. Physicians were therefore expected to be capable of applying both medicinal and mathematical skills in their astrological divinations.

VIII

A RENAISSANCE MAN

Copernicus may have had particular reasons for wishing to remain in Italy to continue his studies. The clear Mediterranean skies would have been far better for continuing his astronomical observations. He would have presumably also valued the contacts he had made at university and his associations with various intellectuals in Italy during his time spent there. He may have had an ulterior motive for readily agreeing to study medicine too.

In the Middle Ages the division between the arts and sciences and various other subjects was not as clearly defined as it is today. Moreover the nature of the subjects in some cases were very different. Medicine for example was closely associated with astrology, which at that time was barely distinguishable from the science of astronomy. In fact astronomy as we know it arose from the astrologers' studies of the heavens.

The astrologer's study of the heavens was for a very different purpose to that of the astronomer, though. Even today some people think that a person's life and the future in general can be predicted by celestial events. It was the astrologer's belief that the sun, moon, planets and stars also affected a person's health, and decided their particular individual constitution, that provided the link with medicine. They thought that the kind of diseases and other ailments people were prone to was influenced by the time of the year they were born and the position of various of the stars, planets and both the sun and moon at their actual moment of birth. It should be remembered maybe that one of Copernicus's professors, Domenico Maria da Novara, had been an astrologer

39. When Nicholas Copernicus finished his studies and returned to Poland it was not to take up his post at Frauenburg but to join his uncle Bishop Lucas Waczenrode at his castle in Heilsberg. Bishop Waczenrode was getting old, his health was not good and he had need of a physician. These were times of great political intrigue and for the Bishop of Ermland assassination was certainly not out of the question; therefore Lucas Waczenrode needed someone he could trust and his own talented nephew Nicholas Copernicus would have been the obvious choice.

but had also taught him much of what he had learnt about astronomy while a student at Bologna University.

The signs of the zodiac represent the constellations of stars one can see in the night sky. These are the apparent groupings of visible stars that had been observed from ancient times. The individual stars are actually at very different distances from the earth and therefore in reality billions of miles apart. Nevertheless, it was by observing how these constellations appeared to move across the sky that allowed sailors to navigate the seas and oceans even before they had compasses. The position of the constellations at different times of the year, combined with their other observations of the sun and moon, also helped the priests of ancient cultures such as Egypt and Mesopotamia to devise calendars by which they could predict seasonal events such as floods, or decide the best time to sow and harvest crops.

'Star gazing' therefore had a long history and it was astrology, and its younger cousin astronomy, that formed the link between mathematics and medicine. This was because mathematics was required to both interpret the observations and make the calculations used in the astrological divinations that were often employed by medieval doctors in their attempts to diagnose a patient's illness. Most people today, on the other hand, would accept that there is no scientific basis for believing that the stars or planets have any influence upon our health.

Copernicus's knowledge of Greek, gained in part from reading ancient texts on astronomy, would have been very useful while a medical student, as many of the manuscripts he would have needed to refer to in his studies would have been in that language. Being a student of canon law he would have also already spoken Latin along with his native Polish and had probably learnt German and Italian too during his student days, so by this time he had quite likely become something of a linguist. He was also by the standards of the time a very well-educated man.

In the Middle Ages it was quite common for those people who were fortunate enough to have a university education to study and become expert in many different subjects and this was certainly the case with Nicholas Copernicus. Indeed, such multi-talented people today are often referred to as Renaissance men or women. But even so he was remarkable for his day. We know for example that he also found time to study art and music as well as all the other subjects already mentioned. In fact the only picture we have of Copernicus is a copy based on a self-portrait that Copernicus painted. He was also later in life both a map-maker and an economist whose ideas on monetary reform are still valid today. Copernicus might not have known it, but many of the skills he had acquired were soon to be put to good use.

40. The Teutonic Order enjoyed the same standing as the Knights Templar and the Hospitallers of St John and had similar privileges, one of which was that they were only answerable to the Pope in Rome. Although originally founded during the Third Crusade, in Jerusalem, as a hospital order specifically dedicated to caring for ailing or wounded Germany knights and soldiery, the nature of their organisation very soon changed and the Teutonic Knights essentially become a group of warrior monks whose objectives were militaristic and self serving. When Acre fell to the Muslims in 1291, the Teutonic Order relocated themselves in Europe and sought lands and power by offering their services as a military force to various monarchs and princes.

When he finally left Italy and returned to Poland it was not to take up his post at Frauenburg. Once again fate played a hand. His uncle Bishop Lucas Waczenrode still had other plans for him. He had recognised maybe that his nephew's broad talents would be largely wasted working as a canon at the cathedral and he arranged for him to join him at his castle at Heilsberg. Furthermore he was in need of a physician. Bishop Waczenrode was getting old and his health was not good. In fact the services of a physician were often only available to the wealthy and it was not unusual for the best doctors to be found at the courts of such people as bishops just as they would have been at the courts of kings or princes. In fact Bishop Waczenrode was to all intents and purposes a prince, having complete autonomy of the duchy of Ermland.

The bishop may have had another more particular reason for choosing his nephew for this post. The duchy of Ermland that he ruled over owed allegiance to the Polish king but was not totally under the bishop's rule. A large part was under the control of the Order of Teutonic Knights who had once acted as mercenaries for one of the Polish princes. Afterwards they had basically commandeered a large area of Ermland and although they had been challenged and defeated by the Polish king at one time they had never given up their territorial ambitions. In fact the Teutonic Order had contrived without success to gain total independence from Polish sovereignty for generations. Bishop Waczenrode's little state acted as an effective buffer between the two antagonists and although as a bishop, with the Church's authority Lucus Waczenrode was the effective ruler of Ermland, which was his diocese; but he nevertheless recognised the wisdom of accepting the Polish king as a nominal overlord. This the Knights vehemently resented as it undermined their own position. Indeed their hatred of the bishop was such that it is said that the Teutonic Knights prayed daily for the bishop's death.

These were thus times of great political intrigue for the bishop of Ermland and assassination was certainly not out of the question; therefore Lucas Waczenrode also needed someone he could trust and his own talented nephew Nicholas Copernicus would have been the obvious choice as his personal physician. Copernicus therefore took up residence with his uncle at Heilsberg Castle which must have been a far more agreeable arrangement than going to Frauenburg to undertake the duties expected of him as a canon at the cathedral.

Although Lucas Waczenrode required Copernicus to serve as his personal physician he appears to have made very good use of his nephew's other abilities too. Apart from anything else he became his uncle's private travelling companion, as well as technical adviser whenever any of his many talents were needed. During this period Copernicus also often served as the bishop's delegate, attending council meetings for example at local towns within the diocese and other official gatherings.

He also acted as his uncle's secretary and this appears to have become his main duty. He was for example responsible for drafting letters to the bishop's many important political and ecclesiastical contacts on his uncle's behalf. This even included corresponding with the Polish king. He must, therefore, have become very informed about the many intrigues in which his uncle was undoubtedly embroiled. At least one of the projects in which Copernicus was involved on his uncle's behalf concerned the troublesome knights. They tried to rid themselves of the Teutonic Order by sending the knights on a crusade. But this scheme came to nothing.

Life at Heilsberg Castle would have certainly been luxurious. Lucas Waczenrode was a prince of the Church and no doubt embellished his residency accordingly. He also had a large retinue. We know from a fifteenth-century ordinance of the castle that there were nine tables occupying the dining hall where the bishop dined with his senior staff and guests. On each occasion mealtimes were conducted with great formality. When the dinner bell was rung everyone else had to wait until the bishop arrived preceded by his hounds that were released to herald his arrival.

Running such an establishment would have required a great many servants, who would have looked after the needs of not only the bishop but his entire entourage as well. Nicholas Copernicus would have enjoyed many privileges as his uncle's secretary and physician that ordinary people during the age could have only dreamt of. Compared to the life of most of the population in medieval Europe, there is little doubt that he led a cosseted existence. Moreover, although he probably had to work hard fulfilling the duties his uncle set him, Copernicus, due to his uncle's influence, was now a canon of the Church and had an income for life.

It is known that Copernicus appreciated everything that his uncle had done for him because he later expressed his feelings in a small book, which he dedicated to the bishop. It was a translation of Greek poetry into Latin and within the book there is a clue to what Copernicus was thinking about at the time. Copernicus had a friend called Rabe, who was a town clerk in the city of Breslau and he had asked this friend to contribute some additional poems for the book. Rabe did so and included a comment referring to Copernicus himself, saying among other things that he, 'explores the rapid course of the moon and the changing movements of the fraternal star' and that, 'he knows how to explore the hidden causes of things'. It is obvious, therefore, that Copernicus was still very much preoccupied with matters of astronomy while living at Heilsberg Castle and appears to have discussed some of his ideas with his friends.

What is important, is that above all else and despite the additional duties imposed on him by his uncle, Copernicus still enjoyed one particular luxury that few others could have afforded in his day – the time to pursue his own interests. Although he made no observations during this period of his life, it was while at Heilsberg Castle that Copernicus began to really develop and refine his new astronomical theory concerning the movements of the planets that was destined to eventually change our ideas about the universe forever. He wrote down a short account of his theories entitled *Little Commentary*, which he circulated privately among his friends. These copies were all hand-written and were therefore limited in number and their distribution was probably confined to people Copernicus thought he could trust.

Copernicus was suggesting that our planet was not the centre of the universe and this was nothing less than revolutionary. He himself was a canon of the Church and yet by putting forward his ideas he was going against Church doctrine. It is no surprise, therefore, that he chose not publish evidence of his research more widely. He may have been afraid of ridicule as much as any physical threat, or sanction, from the Church. Most ordinary people had always accepted that it was the sun moving and not the earth, simply because they could see the sun moving across the sky each day. The idea that it was the earth turning instead, while the sun remained stationary, meant reversing their viewpoint and disregarding what their senses told them. This was difficult for them to comprehend. Nowadays we are far more used to accepting that things are not always what they seem because we have many more ways of examining and testing phenomena, or querying any hypothesis put forward.

41. During April and May 2002, in the very early evening before even the stars had appeared, there was a chance to see each of the planets visible to the naked eye all together. Just as the sun set, if the sky was clear enough and if one was very lucky indeed, even the planet Mercury might occasionally be seen following the sun as it dipped below the horizon. A little higher and further to the east was Venus: often called the evening star and even further to the east one could see Mars, Saturn and Jupiter. These five are very same planets that were known to the ancients before the invention of the telescope.

This phenomenon only occurs when the planets are close to alignment: arranged very nearly one behind the other when observed from the earth and in relationship to the sun. Seen from the Northern Hemisphere in a diagonal line from left to right across the sky, this configuration marks the plane of their planetary orbits around the sun. The orbit of our own planet is very nearly in the same plane and the angle of the paths that the other planets appear to take as they cross the sky is actually indicative of the extent to which the earth is tilted, or inclined, in relationship to the plane of its own orbit.

During the night, as the earth continues to turn, the angle of the observer in relationship to the planetary orbits gradually changes, so that the line they form appears to become less steep as midnight approaches. It is this tilt of the earth, at an angle of 23.5 degrees, that also affects the varying amount of sunlight we receive at different times of the year while the earth moves around the sun; and is responsible for the changing seasons.

IX

QUESTIONING OLD BELIEFS

To understand Copernicus and appreciate what he really achieved as an astronomer involves understanding the world in which he lived and being aware of both the extent and the limitations of people's knowledge at that time. Nicholas Copernicus was born towards the end of the medieval period, which we sometimes refer to as the Middle Ages. It is called the Middle Ages because it is the time between what is known as the Dark Ages, that followed the collapse of the Roman Empire when so much knowledge was lost, and the Renaissance, which means rebirth. The rebirth referred to is the flowering of knowledge, learning and discovery that occurred in the sixteenth and seventeenth centuries. Copernicus played a very important part in this process and as a result of his research and the theories he was to propose we began to change our views and understanding of the universe and the place of our planet in it.

Much of what Copernicus learnt about the earth and the movements of the other planets came from the observations of astronomers who had lived in some cases many centuries before he was born. Eratosthenes – a Greek living in Alexandria, Egypt – had for example performed an experiment to calculate the size of the earth in the third century BC. The method Eratosthenes had used was as ingenious as it was simple. He had noticed while in the town of Syene that on Midsummer's Day the sun shone directly down into a well. He realised that this meant the sun was directly overhead and this gave him an idea about how the size of the earth could be measured. He had placed an upright rod in the ground in his home town of Alexandria and measured its shadow when the sun was at its highest at midday. He then compared this to the measurement taken from an identical rod at midday in Syene that was further to the south. There were no clocks to tell the time as we have today but of course Eratosthenes would have known what the length of the shadow was at midday simply because it would have been at its shortest when the sun had reached its highest point.

By comparing the differing lengths of the shadows at the two locations on the same day he was able to calculate the difference in angle between the two rods in relationship to the sun's rays. This allowed him to work out the amount of curvature of the earth's surface that existed between these two points. He then multiplied the distance between the two rods until the total curvature was enough to form a complete circle. He then knew that the size of the circle would be equal to the total circumference of the earth. The ancient Greeks were also quite sure that the earth was a sphere as this was demonstrated by the shape of the shadow that the earth casts over the moon during a lunar eclipse.

People often think of the sun only being directly overhead on the Equator; but this is not strictly true. The reason why the sun was directly overhead in Syene was because the axis around which the earth rotates is not upright but set at an angle. Eratosthenes knew about this and was also able to correctly estimate the tilt of the earth's axis – which is at approximately 23.5 degrees.

The direction of this angle is maintained as the earth goes around the sun so that the Northern Hemisphere is tilted towards the sun on one side of its orbit to give us summer and is tilted away, to create winter on the opposite side. It is this that creates the seasons, by influencing the angle at which sunlight falls upon the surface of the world in different places at different times of the year. It is hottest when the angle is at its steepest and becomes cooler as the angle decreases because the sunlight is then spread out over a larger area. This is also the reason why the sun appears higher at midday during the summer than in the winter. The opposite happens in the Southern Hemisphere, which is why it is summer there when it is winter in the north and vice versa. By chance, the tilt of the earth on Midsummer's Day in Syene was just enough to allow the sun's rays to shine directly down to the bottom of the well.

Syene would not have been the only place where the sun was overhead at noon on Midsummer's Day. In the Northern Hemisphere the same phenomenon occurs each year during the Summer Solstice at all those locations girdling the earth at the same latitude of 23.5 degrees north: marked by a line circling a modern globe which we refer to as the Tropic of Cancer. In the northern hemisphere the Winter Solstice occurs when the earth is on the opposite side of its orbit around the sun. However, on this day, in the Southern Hemisphere the sun will then be seen directly overhead at all points 23.5 degrees south, marking the Tropic of Capricorn. In fact, the sun can actually only be seen directly overhead at the Equator on two occasions each year: at noon of the Vernal Equinox and six months later at noon of the Autumnal Equinox. It is at these two points during the year that the days and nights are of equal length in both hemispheres.

We also know from the writings of the Greek philosopher Archimedes of Syracuse that he and his friend Aristarchus of Samos, had quite correctly surmised that the earth orbited around the sun and that the stars were an immeasurably large distance from the earth – and this was more than seventeen centuries before the time of Copernicus. Aristarchus was the chief librarian at the famous library at Alexandria as well as being an astronomer. The only surviving work of Aristarchus is a short treatise entitled *On the Sizes and Distances of the Sun and Moon*. The values he obtained were inaccurate, due to faulty observations, but one of the methods he devised for measuring the distance of the moon was employed by a later astronomer called Hipparchus.

Less than a century later Hipparchus managed to measure the distance of the moon from the earth with remarkable accuracy, obtaining a value equal to sixty earth radii, or in other words thirty times the diameter of the earth. He did this by comparing the results of two separate observations taken during an eclipse of the sun. One was taken at Syene where a total eclipse was seen and the other in Alexandria, where only a partial eclipse was recorded. By assessing the percentage of the sun that was covered by the moon during the partial eclipse in Alexandria, and measuring the distance between there and Syene, Hipparchus could work out the distance of the moon from earth using simple trigonometry. This is because the amount of the sun visible at the second location told him the difference in the angles at which the moon had been viewed by the two observers, allowing him to calculate the moon's parallax. This is essentially very similar to the type of experiment described earlier, which Copernicus did with his teacher Professor Domenico Maria da Novara, while at Bologna University – although on that occasion the calculations of Copernicus and his tutor were not dependent upon an eclipse occurring.

Much of this knowledge was lost during the Dark Ages that followed the fall of the old Roman Empire but by the time Copernicus was born towards the end of the Middle Ages some of this ancient wisdom had been regained following contact with Islamic cultures that had preserved old manuscripts

Summer in the Northern Hemisphere

Autumn in the Northern Hemisphere

Winter in the Northern Hemisphere

Spring in the Northern Hemisphere

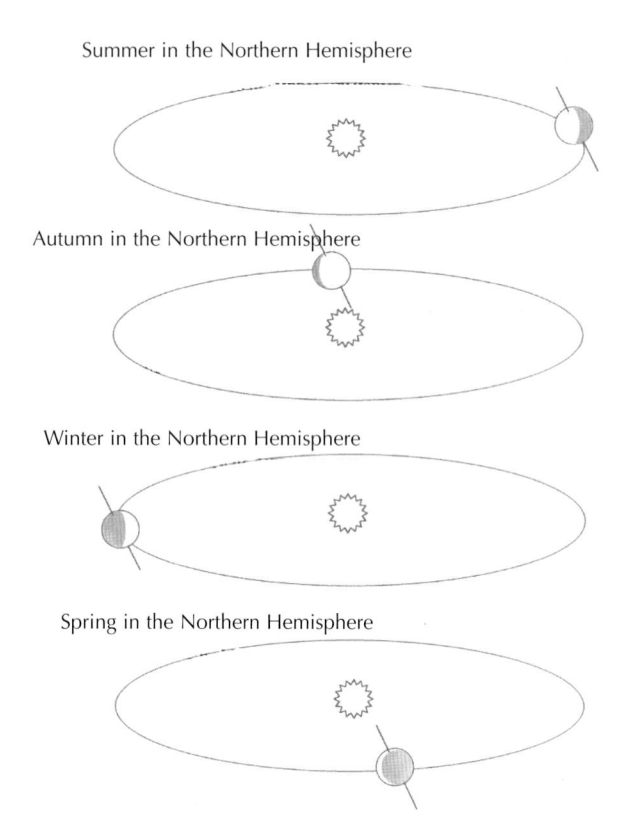

42. The changing seasons are due to the way in which the earth is tilted, or inclined to the plane of the ecliptic, which is the line defining both the apparent path the sun takes across the sky and the earth's actual orbit around the sun. The inclination of the earth does not change as it orbits the sun; only its orientation. It is summer in the Northern Hemisphere when the North Pole is tilted towards the sun and winter when it is tilted away; whereas at the same time in the Southern Hemisphere it is winter and summer respectively. Between these seasons there is spring and autumn, during which time the sun is briefly directly over the Equator at the spring and autumn equinoxes.

acquired during their conquests of the lands which were once part of Greek and Roman civilisations. It was this that was to gradually lead to the Renaissance, or rebirth, of knowledge in Europe that had already begun within Copernicus's lifetime. The information, though, emerging from these sources was piecemeal and there were still areas of great ignorance. It had only been recently accepted, for example, by most people in Europe that the earth really was a sphere; although this was only really confirmed once when sailors such as Magellan had sailed around the world. Even then people did not realise that the earth turned – it was thought that the movement of the sun, moon and stars was because they all revolved around the earth: whereas in fact only the moon did.

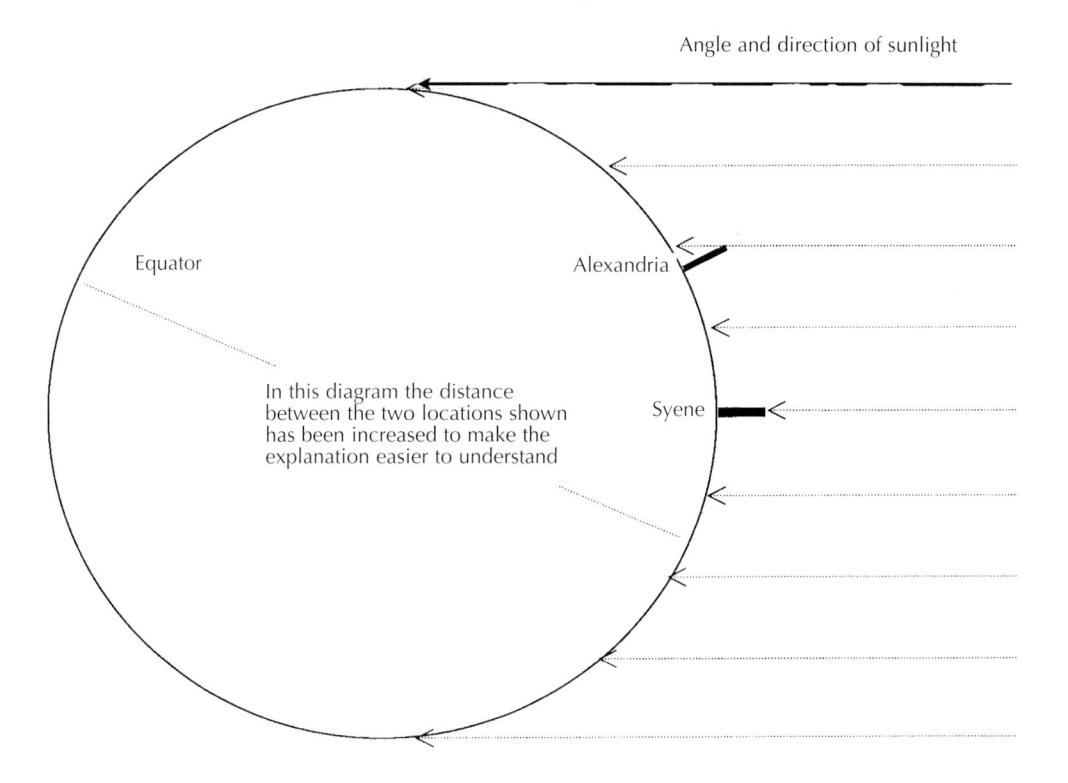

[Diagram: Eratosthenes' method showing Earth as a circle with Equator, Alexandria, and Syene marked, with arrows indicating "Angle and direction of sunlight" hitting the Earth at different points. Note on diagram: "In this diagram the distance between the two locations shown has been increased to make the explanation easier to understand"]

43. The Greek astronomer Eratosthenes was able to calculate the size of the earth with remarkable accuracy even in the second century BC. The diagram shows how the sun's rays hit earth at different angles on its circumference. Eratosthenes measured these angles using a rod at midday in two places of a known distance apart. He then used mathematical calculations to work out the total circumference of the earth. This relatively simple process brought them an important advance in their knowledge about the nature of the earth.

Astronomers in the past may have assumed that Mercury and Venus were nearer to the sun than the earth because they took less than a year to complete their orbits whereas all the other planets then known took longer than a year. There was, though, uncertainty about which was the nearer to the earth. It seemed obvious that the orbit of Mercury was smaller than that of Venus simply because the positions it occupied in the sky at elongation, when on opposite sides of its orbital path, were closer than those of Venus. This meant that if the sun, moon and all the other planets went around the earth, as many people once thought, Mercury had to be nearer than Venus. Conversely though, it also meant that if the sun was at the centre, Venus with its larger orbit must be nearer to the earth whilst Mercury was nearer to the sun. This we now know is the true situation.

Altitude of the midday sun at the Summer Solstice in the Northern Hemisphere

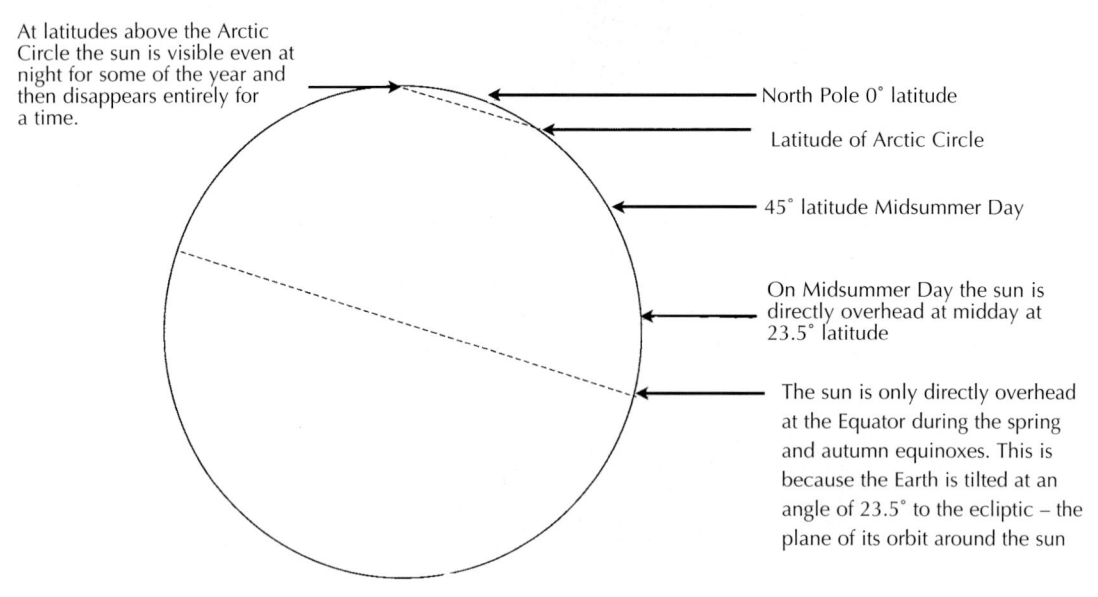

At latitudes above the Arctic Circle the sun is visible even at night for some of the year and then disappears entirely for a time.

North Pole 0° latitude

Latitude of Arctic Circle

45° latitude Midsummer Day

On Midsummer Day the sun is directly overhead at midday at 23.5° latitude

The sun is only directly overhead at the Equator during the spring and autumn equinoxes. This is because the Earth is tilted at an angle of 23.5° to the ecliptic – the plane of its orbit around the sun

Altitude of the midday sun at the Winter Solstice in the Northern Hemisphere

There are times during the year when the sun for some periods is no longer visible from regions within the Arctic Circle.

Latitude of the Arctic Circle

45° latitude Midwinter Day

Midwinter Day at 23.5° latitude

The angle at which sulight reaches earth at midday on the Equator at Midwinter Day

The sun is directly overhead at midday at 23.5° latitude in the Southern Hemisphere when it is Midwinter Day in the Northern Hemisphere

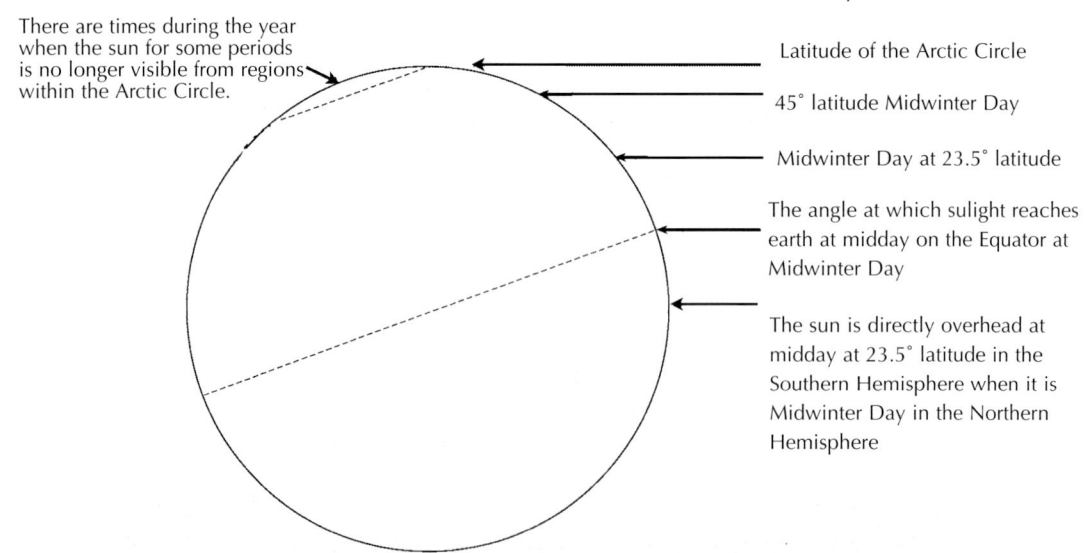

44. Using an astrolabe or quadrant it was possible to find one's latitude by measuring the height of the sun at midday. But due to the changes seasonally seen in the apparent position of the sun, mariners in Copernicus's day would normally have needed tables, accurately predicting the extent to which the elevation of the midday sun at different latitudes would vary during the year. Producing such tables was one of the tasks that had often been undertaken by astronomers over the centuries.

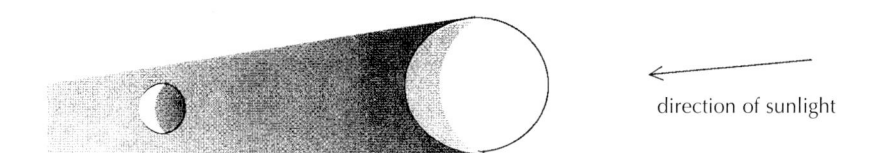

In the third century BC the Greek astronomer Aristarchus tried to discover how large the moon was in comparison to the earth by measuring the size of the earth's shadow as it crossed the face of the moon during a lunar eclipse. He estimated that the earth had a diameter twice as big as that of the moon, because of the time the moon remained in the earth's shadow indicated that the shadow was twice as large as the diameter of the moon. Aristarchus did not realise that because of the sun's great size the size of the earth's shadow decreases with distance, to form a cone that only covers two-thirds of the area it would otherwise cover. The diameter of the moon is in fact only a third of the earth's.

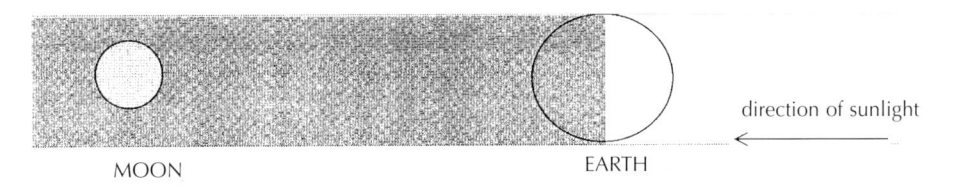

The diagram above shows the earth's shadow as Aristarchus imagined it to be whereas the diagram below shows that in reality the shadow cast by the earth forms a cone.

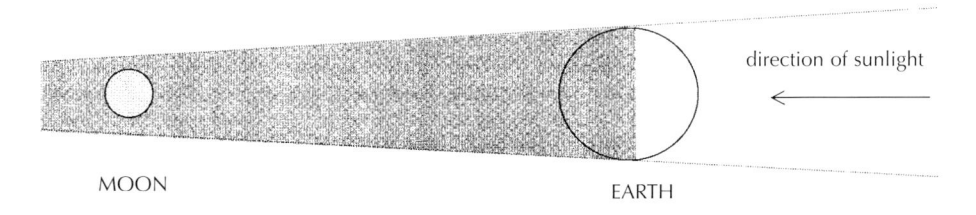

Eratosthenes – a Greek living in Alexandria, Egypt – had already performed an experiment to calculate the size of the earth in the third century BC. By comparing the angle of the shadows cast by the sun during the summer solstice at midday in two widely separate locations. Therefore with this information in theory Aristarchus could have calculated the size of the moon if his experiment had not been flawed by a perfectly simple and understandable misconception.

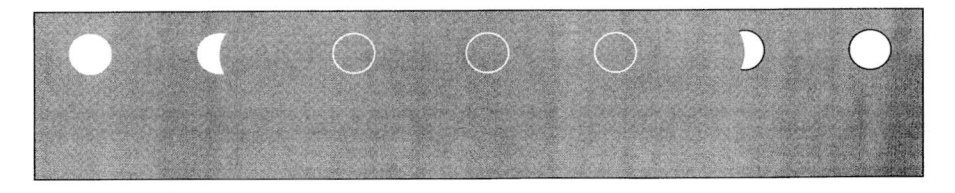

For half the entire duration of the eclipse the moon is totally obscured by the earth's shadow, which indicates that the shadow cast by the earth must be twice as wide as the moon's face, even though the earth's diameter is actually three times as large as the moon.

What observer A would have seen at Syene

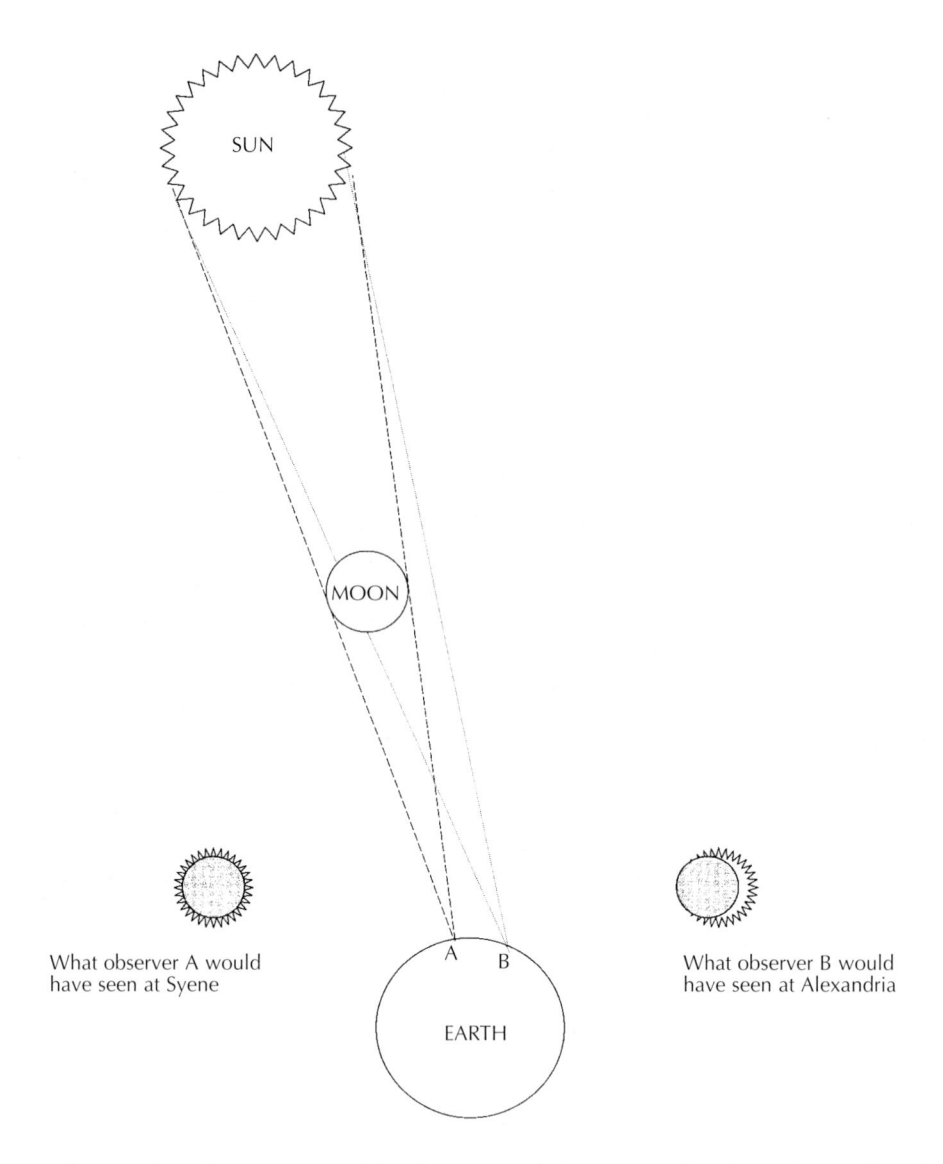

A B

What observer B would have seen at Alexandria

EARTH

46. When a solar eclipse occurs, a full eclipse can only be witnessed in those places where the moon's passage across the sky results in it coming directly between an observer and the sun. Only a partial eclipse can be seen by those people who are on the edge of the area of shadow produced as the moon's path takes it in front of the sun.

In the second century BC the astronomer Hipparchus used this phenomenon to conduct an experiment designed to determine the distance of the moon from the earth. His observations showed that when there was a total eclipse in Syene an observer in Alexandria would only see a partial eclipse, with one-fifth of the sun remaining visible. Using this information – and calculating the angular difference involved – he was able to roughly estimate how far the moon was from the earth. This he believed to be equal to approximately fifty-nine earth diameters, whereas the distance on average is actually closer to fifty-seven diameters. Nevertheless it was an remarkably good result considering the methods he had to use and his lack of instrumentation.

X= position of observer

47. Two astronomers from ancient Greece, Aristarchus and Hipparchus, each attempted to discover how far the moon was from the earth. Once the distance of the moon from the earth had been calculated it was also theoretically possible to assess the distance of the sun from the earth using simple geometry. The method used is based upon the assumption that when the moon is half illuminated by the sun, the earth and the sun are at right angles to the moon: forming a right angle triangle as in the diagram below. Therefore, by also measuring the angle between the sun and the moon when they are both visible in the sky, and knowing the distance to the moon, one has all the necessary information to work out the two unknown distances representing the other two sides of the triangle. The one surviving work of Aristarchus is a short treatise entitled *On the Sizes and Distances of the Sun and Moon*. However, the values he obtained were incorrect. This meant that his estimate of the moon's distance from the earth was not accurate enough to help estimate the distance to the sun, nor was his measurement of the angle between the sun and the moon precise enough to be used in calculating the vast distances involved: as the distances increase any error in the calculations used are compounded. Aristarchus believed that the angle between the sun and moon, when the moon was half full, was 87 degrees, whereas in reality it is 89 degrees, fifty minutes. It was only when much more reliable equipment became available, over 2,000 years later — and long after the time of Copernicus – that the method described could be effectively employed to obtain the correct distances. We now know that the moon is approximately a quarter of a million miles from the earth and that the sun is about 93.25 million miles away, a distance that is nearly twenty times greater than the one obtained by Aristarchus.

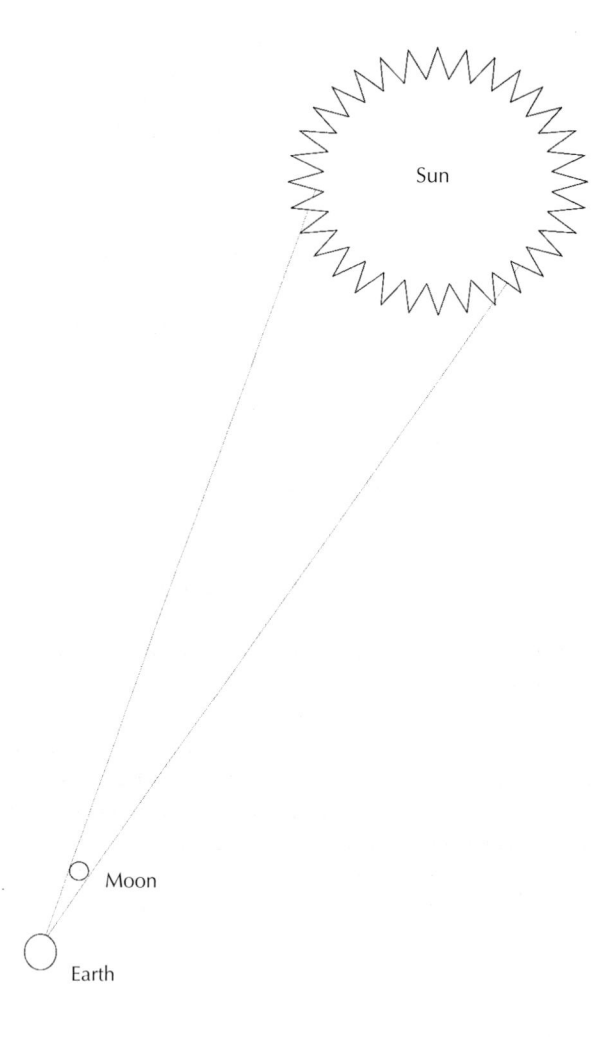

48. The sun and moon appear to have the same diameter when seen in the sky. Nevertheless it was known from very ancient times that the sun must be bigger than the moon because during a solar eclipse it is hidden by the moon and therefore must be further away: which means that the sun has to be larger just to appear the same size.

Many astronomers in the past, who believed that they already knew the distance and size of the moon and the true distance of the sun, have tried to use these facts to also assess the sun's size.

Because the sun and moon appeared the same size they quite rightly assumed that, if for example the sun was twenty times further away than the moon, it must be twenty times larger too. This was the distance estimated by the Greek astronomer Aristarchus. The mathematics was correct but the information was wrong. All the ancient astronomers did in fact underestimate the size and distance. The sun is actually nearly 400 times further away from the earth than the moon.

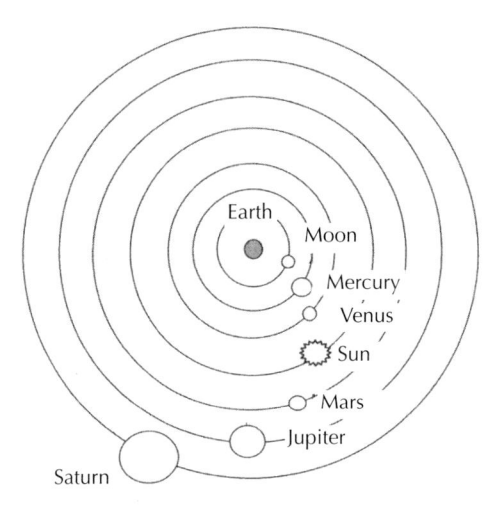

49. Ptolemy thought that the earth was at the centre of the cosmos: proposing a model of the cosmos that is referred to as a geocentric system. His reasons for this were based upon quite simple direct observations rather than theoretical principles. He knew that the moon was the nearest body to the earth because it occasionally came between the earth and the sun during an eclipse, as well as sometimes passing in front of the other planets. He had also observed that Venus and Mercury sometimes crossed in front of the sun and therefore assumed that they must have orbits between the earth and the sun: although he believed that as Mercury had the smallest orbit it had to be the nearest of these two planets to the Earth. Using much the same reasoning he placed the other planets of Mars, Jupiter and Saturn beyond the sun because they never came between the sun and the earth. Similar ideas had been suggested by earlier astronomers; but the system Ptolemy proposed, although logical, was wrong. Copernicus on the other hand realised that there were other ways in which the same observations could be interpreted.

People knew too that the moon was nearer than the sun as a result of observing the various eclipses of the sun that occurred and would have assumed that the moon was nearer than the planets not only because it appeared bigger, but also because it too sometimes passed in front of a planet. But in the Middle Ages it was thought that the sun too circled the world in an orbit between the inner planets and three other planets, Mars, Jupiter and Saturn, that were in fact the only others that could be seen before the invention of the telescope.

The stars were accepted as being the farthest of the celestial bodies from the earth, circling round outside the orbit of Saturn as it could be seen that the planets sometimes crossed in front of the stars; although people of that time had no concept of the vast distances that were involved and this influenced many of their ideas about the nature of the universe. When one

50. Here one can see how, as the earth overtakes it, the planet Mars seems to temporarily reverse its movement across the night sky, while passing through the constellations. The broken line marks the ecliptic, the apparent path taken by the sun as it crosses the sky. This also represents the approximate position of the slightly varied planes in which each of the planets orbit the sun. We can see how closely the orbit of Mars, being in nearly the same plane as the ecliptic, follows a very similar path to that apparently taken by the sun. We now know that the sun only appears to move in this way because the earth is turning and that this merely coincides with the movements of the other planets due to the fact that the orbit of the earth and the orbits of the other planets known at the time of Copernicus are in a very similar plane. It was this phenomenon that led most early astronomers to initially assume that the sun as well as the moon, along with all the other planets, orbited the earth.

object passes in front of another in this way it is called an occultation. One of the earliest astronomical observations that we have a record of Copernicus doing was when he observed the occultation of the star *Aldebaran* by the moon which he made on 9 March 1497 while studying at the University of Bologna in Italy.

It has already been mentioned that that Archimedes and his friend Aristarchus both believed that the earth orbited around the sun. The trouble was that the philosopher Aristotle and an ancient Egyptian astronomer, Ptolemy of Alexandria, both thought that the earth was the centre of the universe. Although Ptolemy had lived centuries before, people tended to still refer to his model of the universe because they believed that his ideas were correct – but they were not. This caused a lot of problems when astronomers tried to explain the movements of the sun and stars – and especially the movement of the planets. In fact Ptolemy was wrong about nearly everything his astronomical theories had tried to explain.

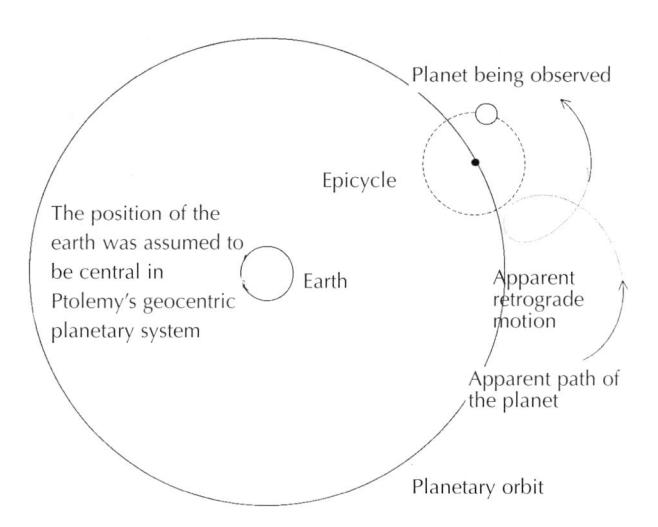

51. Thanks to Copernicus we now know that the apparent retrograde motion of the planets can be explained by the differences in the time each planet takes to orbit around the sun. In medieval Europe many astronomers still believed that the sun orbited the earth and therefore sought an alternative explanation. Most accepted the theory of the astronomer Ptolemy, of second-century Alexandria, who had suggested that the planets might also revolve around a central point, in smaller circles called epicycles while travelling along their orbit. Because this seemed to provide an explanation for retrograde motion this idea was accepted for centuries even though it was completely wrong.

One of the few things Ptolemy was right about was when he said that the moon went around the earth. But he also thought that the sun went around the earth and that the stars did too. It is easy to understand why Ptolemy believed that the stars revolved around the earth because they seem to move across the sky every night from east to west: in fact just as the sun seems to do during the day. He knew too that the planets of Mercury and Venus came between the earth and the sun and that the other visible planets of Mars, Jupiter and Saturn did not. Ptolemy also quite wrongly thought that Mercury was the nearest planet to the earth, rather than Venus. This was because Mercury with the smaller orbit was the fastest and so in an earth-centred cosmos it would have to be the nearest planet, as explained earlier.

Explaining the actual movements of the planets using Ptolemy's ideas was much more difficult. The word planet is actually from ancient Greek, meaning wanderer. This is because of the strange movements each planet appears to make in the night sky.

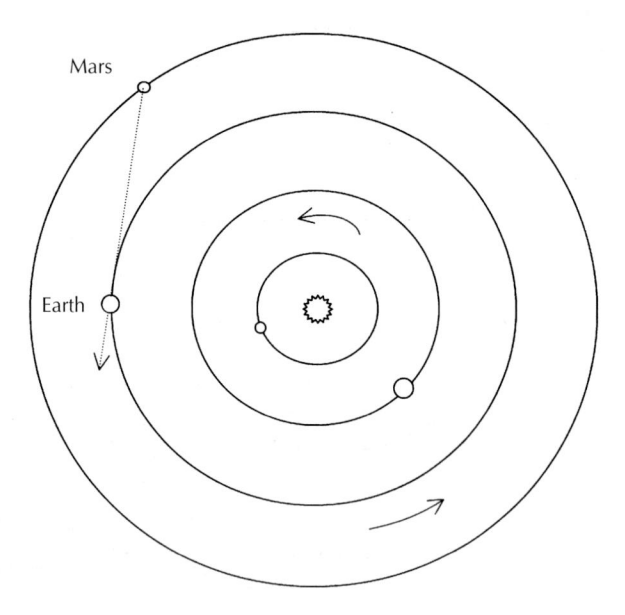

52. The stars never vary the pace of their apparent movements each night; but when observed from the earth the other planets will appear to travel at different speeds as their relationship to the earth changes during their passage around the sun, even though their speeds are practically constant. At the point shown in the diagram above for example the planet Mars will appear practically stationary as the earth is moving directly away from it. A similar thing happens when the earth is moving directly towards Mars, as seen below. It was these changes in apparent speed that gave early astronomers a clue to the planets' movements and their relationships to the earth.

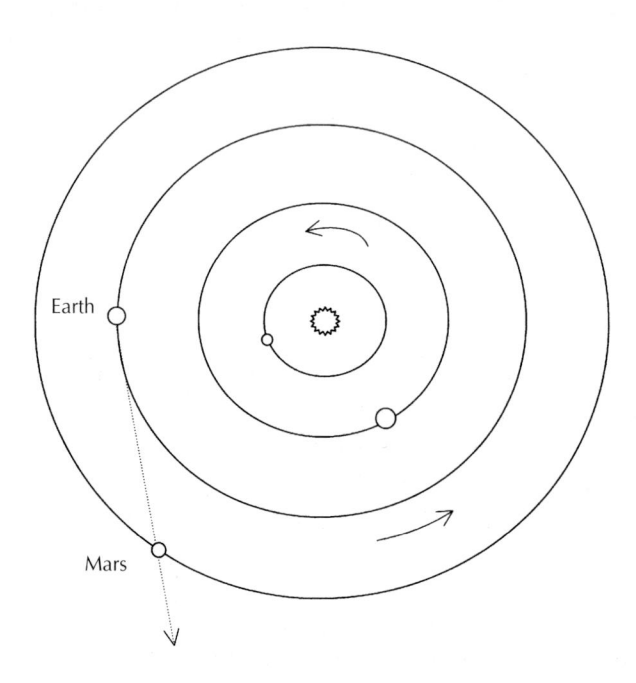

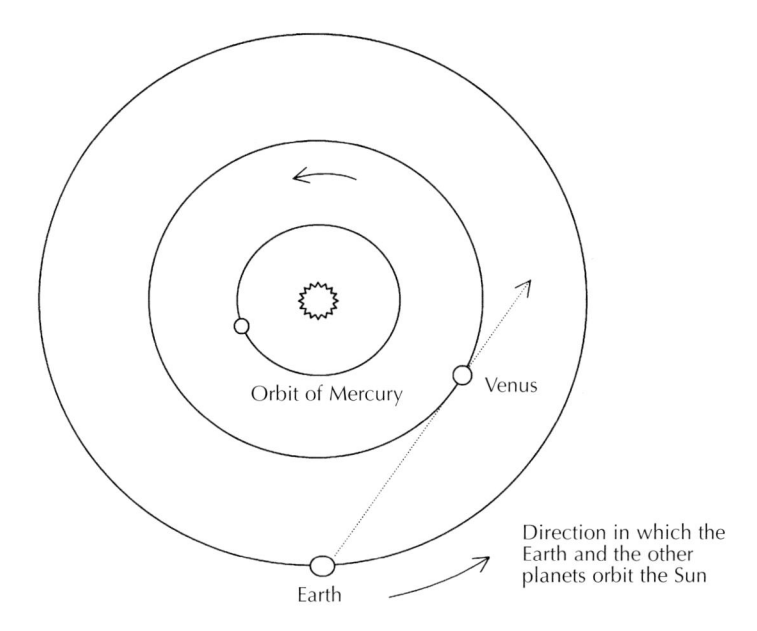

53. By observing events, such as those depicted, ancient astronomers were able to determine that the planets followed circular paths.

The dotted, arrowed line above indicates the point at which an inferior (inner) planet, such as Venus, will appear to become stationary for a period of time as it moves directly away from the earth, while orbiting the sun. A similar phenomenon is also seen to occur when an inferior planet is moving directly towards earth as is illustrated below.

An inferior planet is said to be at its greatest elongation when the relationship between it and the earth makes it appear to have become practically stationary because this occurs at a time when the planet appears furthest from the sun.

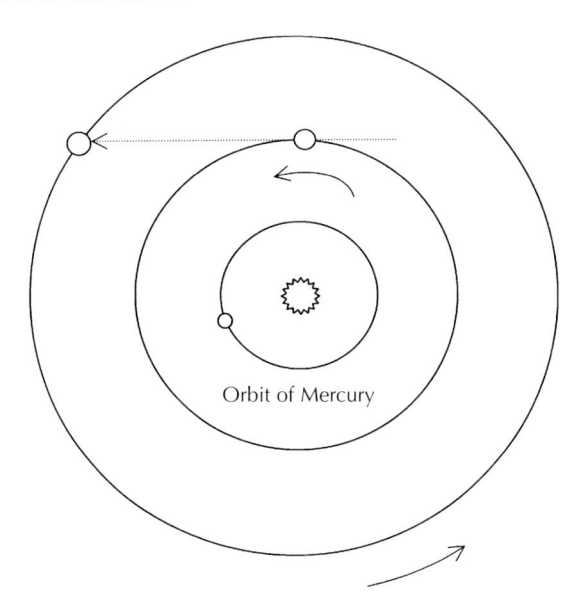

54. When one watches one of the outer planets like Mars over a period of time such as a year, one can see that it does not always move in a straight line across the sky as the stars do, but instead follows a path with a number of loops. This makes the planet appear to be sometimes even going backwards on its tracks. This motion is referred to as retrogression.

The ancient Greek astronomer Ptolemy had sought to explain this phenomena by inventing the idea of epicycles, suggesting that the planets each followed smaller circular paths in addition to their main orbit, which he thought was around the earth, which Ptolemy erroneously believed was at the centre of the cosmos.

Copernicus was not content with Ptolemy's view of the cosmos, and the apparent retrogression of the planets gave him a clue as to what might really be happening. Copernicus noticed that the looping paths of the outer planets, in which Mars, Jupiter and Saturn seemed to go backwards, all had one thing in common.

These movements all followed repetitive, annual sequences. Mars, the nearest of the outer planets to the earth, appeared to reverse its motion twice a year. The next planet, Jupiter, did so twelve times in a year and Saturn, the furthest planet visible to astronomers at that time, had twenty-nine retrogressions.

Copernicus therefore reasoned that as the apparent retrogression of the outer planets seemed to be an annual phenomenon they must be related in some way to the earth year. In fact this feature of planetary motion is due to the difference in time it takes each planet to complete an orbit of the sun, with the planets furthest away with the largest orbits taking longer than those nearer. These annual repetitions of retrograde motion may have provided one of the most persuasive pieces of evidence Copernicus had in support of a heliocentric or 'sun-centred' model for the cosmos.

If you watch a planet for a very long time, through a whole year for example, it does not move in a straight line across the sky like the stars do, but follows a path of loops instead. Some planets make more loops than others and different planets make different size loops. Most planets look exactly the same as stars without a telescope but it was because of the odd behaviour that people knew long ago, before telescopes were ever invented, that planets were different to stars.

The looping paths of the planets made them appear to be sometimes going backwards on their tracks. This motion was referred to as retrogression. Ptolemy's imagined that this was because the planets also each followed smaller circular paths which he called epicycles, in addition to their main orbit as they revolved around the earth, known then as a deferent. This idea enabled Ptolemy to explain the apparent movements of the planets and even predict their paths with reasonable accuracy. For this reason his theories were believed to be correct for 1,000 years, even though they were wrong.

Epicycles were also used by Ptolemy to try and explain the uneven movement of planets. Whereas the stars move steadily together across the sky at the same pace night after night; the planets do not. Their movements are far more erratic. Sometimes they appear to speed up and move further across the night sky within a set length of time and at other times slow down and even stop. The real reason is because the planets follow an orbit around the sun that sometimes means they are moving away on a path that is in direct line with the earth or coming straight towards it. Under these circumstances no movement can be detected even though their real speed has not significantly changed. This phenomenon is referred to as elongation. Alternately, when planets are seen passing directly across the sky they seem to move faster. None of this could be properly understood in Copernicus's day because people thought that the planets, along with the sun, moon and stars revolved around the earth.

Copernicus was not content with Ptolemy's view of the cosmos. The clue was in the apparent retrogression, or reversal of direction, that was occasionally seen in the path the planets took as they crossed the heavens, which Ptolemy had sought to explain by inventing the idea of epicycles. Copernicus noticed that the looping paths of the outer planets, in which Mars, Jupiter and Saturn seemed to temporarily go backwards, had one thing in common. These movements all followed, repetitive annual sequences. Mars, the nearest of the outer planets to the earth, appeared to reverse its motion twice a year. The next planet, Jupiter did so twelve times in a year and Saturn, the furthest planet visible to astronomers at that time, had twenty-nine regressions. He, therefore, reasoned that as the apparent regression of the outer planets seemed to be an

The outer planet will appear to make a loop here

The lines numbered 1-10 are lines of sight when viewing the position of an outer planet against the stars as seen from the earth

Direction of both planetary orbits

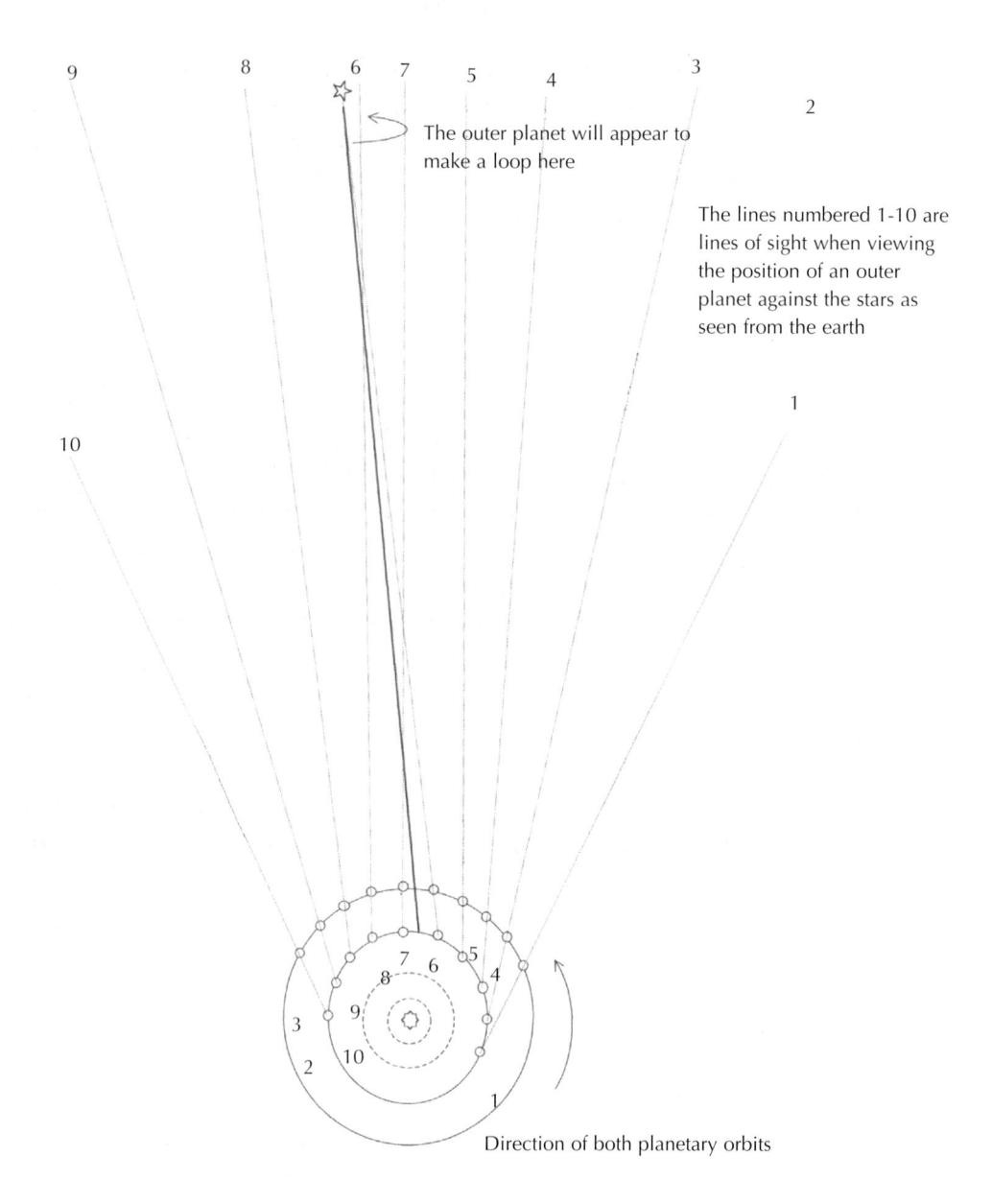

55. One can see by comparing the shortening distance between the dotted lines of sight how the movement of an outer planet such as Mars when viewed from the earth would seem to become slower as it approaches the star shown on the extreme left of the page. Nevertheless, up until position 5 the outer planet (when seen from the earth) still appears to be approaching the star. But when the earth is midway between 5 and 6 (as indicated by the solid line) the outer planet would briefly appear stationary against the background of the stars. It would then seem to go into reverse when observed from the earth and for a short time, to move westward away from the star. But by the time the inner planet had reached position 7 the outer planet, its path having made a short loop, would once more appear to be moving eastwards as before.

annual phenomenon they must be related in some way to the earth year. In fact this feature of planetary motion is due to the difference in the time it takes each planet to complete an orbit of the sun, with the planets furthest away with the largest orbits taking longer than those nearer. These annual repetitions of retrograde motion may have provided one of the most persuasive pieces of evidence Copernicus had in support of a heliocentric or 'sun-centred' model for the cosmos.

The truth, as we now know, is that stars do not revolve around the earth at all; nor does the sun. What is really happening is that the earth turns once every twenty-four hours and it is this that makes it look as if the stars and sun are going around the world. People in the Middle Ages did not understand that it is this that really decides the length of a day and causes the changes we see from day to night, as they thought the sun orbited around the earth; whereas in reality it is the movement of the earth as it rotates that causes these changes. We now accept that when our side of the earth is facing the sun it is day-time and when our own side of the earth is turned away from the sun it is night-time; but in the Middle Ages it was generally believed that the earth was completely stationary. Neither did they realise that the earth orbits around the sun.

In medieval times many people preferred to believe that the earth was the centre of the universe because this fitted with their ideas about humankind's special place as being central to the scheme of things in the biblical story of the creation. The theories of Ptolemy that placed the earth at the centre of the universe seemed to completely support this idea and the Church in particular wanted to believe this because it fitted with their own belief about the special importance of our world in a universe created by God. Therefore many churchmen thought that any ideas suggesting that the earth's place in the universe was not special may cause people to question other beliefs upon which Christianity was based. There was, in fact, frequent intolerance and prejudice among certain sectors of the Church towards anyone who raised doubts about any aspect of their religious teaching. The Old Testament even describes the earth as 'un-moving', which made the new ideas of Copernicus even more difficult to accept for many people.

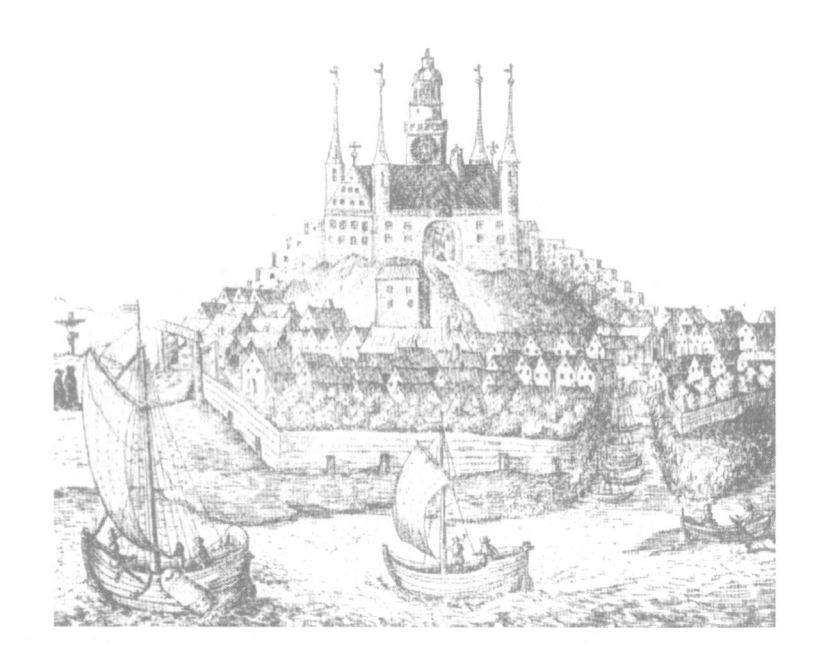

56. Frauenburg was in the duchy of Ermland and the cathedral that dominated the city was the official seat of Copernicus's uncle Bishop Waczenrode who ruled over the duchy with the powers of a prince. Copernicus was officially a canon of the cathedral and it is here that he dwelt in the latter part of his life. The city was strategically quite important and well fortified in Copernicus' day, having suffered several times during various skirmishes that occurred periodically between the Order of Teutonic Knights and their nominal overlords – the Polish monarchy. One of the most serious conflicts occurred while Copernicus was actually in residence, when the city was being besieged by the Teutonic Order.

X

THE CANON OF FRAUENBURG

Despite his many travels and varied duties over the years we should not forget that Copernicus was actually still officially a canon of the cathedral in the town of Frauenburg. It was probably only because his uncle had found his services so indispensable and in particular needed him as his personal physician in later years, that Copernicus was excused his other duties. But this was all to change when on 29 March 1512 Bishop Lucas Waczenrode died.

The two men had visited Kraków to celebrate the wedding of King Sigismund, but the bishop had begun the return journey to Heilsberg without his nephew and on the way back was suddenly taken ill. He had previously been in perfectly good health. Although the symptoms could be interpreted as being nothing more sinister than food poisoning, this unexpected turn of events was regarded with some suspicion. His condition deteriorated rapidly and as the weather was cold he was taken to the nearest place of shelter which was Toruń, his birth-place. But he was a man already in his mid-sixties and he never recovered. It is impossible to say whether or not Copernicus could have saved his uncle and we do not know why he had not travelled with his uncle on that occasion; but at the very time when Lucus Waczenrode most needed his services as a physician fate had decreed that his nephew should be elsewhere.

Now, with his duties as an aide and personal physician to his uncle ended, Nicholas left Heilsberg and moved back to Frauenburg to finally take up his post as a canon, living for most of the time along with the other resident clerics within the precincts of the cathedral.

57. Leprosy was common in Europe throughout the Middle Ages. It was known that the infection could be passed on via open wounds and lepers had to warn people of their condition by ringing a bell: therefore it was understood that it was contagious. Special leper hospices existed that were run by many monasteries and convents; but no cure was available.

There were sixteen canons in total at Frauenburg Cathedral including both Nicholas Copernicus and his older brother Andreas. Although they had all taken vows of obedience and were meant to lead a celibate existence and serve the Church to which they belonged, this by no means meant that their lives can be equated with that of ordinary monks, who had to give up all their worldly goods and possessions. The canons were allowed to carry arms and were expected to possess several horses as well as maintaining servants. By comparison to ordinary monks the canons of Frauenburg and similar establishments were without doubt privileged and pampered. Their prescribed duties were also very light.

To begin with Nicholas would have had the company of his older brother Andreas, but their time together was brief, as unfortunately his brother had become ill. The precise nature of his illness is not known for certain but what we do know is that it caused sufficient physical disfiguration and fear of contamination for the other canons to ask him to leave the cathedral. It used to be thought that he may have contracted leprosy.

Leprosy is basically a disease of the nervous system. The virus damages the nerve endings and the patient being unable to feel properly tends to injure various parts of their body, which results in open sores and ulcers occurring

58. Despite his many travels and varied duties over the years Copernicus was still officially a canon of Frauenburg Cathedral. Copernicus was initially excused his duties as a canon because his uncle the bishop had employed him as his secretary and his personal physician. But this was all to change when Bishop Lucas Waczenrode died. Nicholas then left the bishop's palace at Heilsberg and moved back to Frauenburg to finally take up his post as a canon, living for most of the time along with the other resident clerics within the precincts of the cathedral. It was while living in Frauenburg that he completed the work on his theory of a sun-centred planetary system.

and additional infections that further increase the damage until fingers and toes are lost as well as other disfigurements. In the penultimate stages the patient's face is also affected by the virus, becoming so distorted that the condition is actually referred to as 'Lion Face' due to the similarity of appearance with that animal. Without treatment an amelioration of these symptoms is normally only seen shortly before the patient dies – and in the Middle Ages no such treatment was available. It is not at all certain though that Andreas had leprosy. Although his condition was described as '*lepra*' this is a word that had several connotations during the Middle Ages which may better be translated into today's common idiom as '*the pox*'. This term was actually used to refer to several diseases displaying similar symptoms such as ulcers and open sores. From what little we know about his condition, and the general abhorrence obviously felt by the chapter regarding his presence, it is quite possible that Andreas was instead suffering from syphilis.

We know from a number of sources that syphilis was definitely present in medieval Europe during the fifteenth century. Indeed it may have been there for centuries but in a relatively mild form that caused little lasting harm and to which the population over many generations had acquired some resistance. A very similar scenario seems to have existed in the New World too and it is now suspected that sexual interactions between European explorers and the natives of the New World meant that an interchange of the different forms of the virus occurred, which resulted in new forms being introduced into both populations to which they had little resistance. The outcome was that each of the two unfamiliar strains of the virus were both manifesting themselves in far more virulent forms either side of the Atlantic. In its new guise syphilis, which may have previously only presented as a mild infection, may not have been recognised for what it was. Many physicians may have misdiagnosed it as a form of leprosy due to the ulcers and pustules patients suffered, whereas to others it could have appeared to have been an entirely new disease. In fact, to all intents and purposes it may as well have been, as there was no natural immunity and it soon took on alarming proportions with each part of the Old World blaming the other for introducing it. Some thought that it had originated in Italy but it was also referred to as the French Disease, while others believed it came from Arabia.

Columbus had returned from his first voyage to the Americas in 1492 and a document published in 1495 makes it clear that some form of syphilis was widespread in Europe by that time. Whether or not this was a strain introduced from the New World it is impossible to say but it seems likely that the new virulent form could quite easily have already been present in Rome when the two brothers, Nicholas and Andreas Copernicus, were there during the centennial celebrations of 1500. We know that the two young men lived it up and regardless of the vows they had taken, the one of celibacy was frequently flouted by all *echelons* of the clergy during the period in which they lived. Even the pope incumbent while they were in Rome, Rodrigo Borgia, had a mistress and indeed could be regarded as a family man having sired several children – the notorious Cesare Borgia being one of them.

Syphilis and leprosy were just two of the scourges of medieval Europe and regardless of which of these diseases Andreas was suffering from there was no suitable treatment or remedy available at that time for either. This of course did not stop people seeking a cure. With Andreas the disease must have been already well advanced when Nicholas rejoined the chapter following the death of their uncle the bishop, as by 1512 the other canons had tried to instigate his removal from their small community. He did not go willingly and even after obtaining his reluctant agreement to leave, he hung around the town causing everyone, including his own brother, acute embarrassment for another couple of months.

Finally when all attempts to rid him of the disease had failed and the chapter could no longer bear his hideous presence in their midst, Andreas

59. This was a time of discovery, as Columbus voyaged to the Americas in 1492 and world travel became more common as people sailed from the New World to the Old World in search of new opportunites. Thus it is not surprising that disease were transmitted across the oceans and brought to major cities, such as Rome.

departed. His destination was Rome where he had probably first contracted the disease. He may have seen reason and chose to leave to avoid passing on the infection to his brother canons, or he may have also left in the hope of finding a cure elsewhere. We do hear of him again briefly while he is in Rome, becoming involved in Church politics. Despite his illness and unpleasant appearance he was apparently accepted more easily there; but then silence. What his fate was from this point on we do not know; he disappears from Copernicus's life never to be heard of again. It is thought that he died only a few years after he had left.

In fact being medically trained, one of Copernicus's duties would have been to serve as a physician to the cathedral and the local community. There was a strong link between religion and medicine in the Middle Ages. It was often assumed for example that diseases of the body resulted from the sins of the patient – a kind of sickness of the soul. This may go some way towards explaining the chapter's abhorrence of Andreas's condition and their embarrassment as fellow canons. Nowadays we might refer to certain diseases in a similar way when talking of psychosomatic illnesses. But in the Middle Ages people believed that most ailments that were not due to an injury, or could not be easily explained in some other way, were caused by the spiritual shortcomings of the afflicted person. Many people, therefore, sought relief from their ills through meditation, prayer, pilgrimages, and other non-medical methods. The ministry of the clergy was therefore thought to be every bit as important as that of the physician. This was probably a major consideration when the young Copernicus was originally given permission to continue his later studies at university on the proviso that he pursued a course in medicine.

Some of the practices employed by the medieval physician were not dissimilar to those of today: as with his modern counterpart the physician would usually take a patients pulse to assess his health just as the modern-day doctor does. Blood samples too were used to help in diagnosis. But one of the best diagnostic tools available to the medieval doctor was a uroscopy, in which the colour of the patient's urine was examined to determine what ailed the patient and decide what treatment was required. There were charts produced showing the various colours of urine which the doctor could refer to in order to identify an illness and determine the possible cause.

A range of treatments were employed by medieval doctors, that could simply mean administering laxatives and diuretics or prescribing hot baths; but might equally well involve the cauterisation of a wound, or an infected area of the body, fumigation, the administration of herbal remedies and blood-letting. As with the Greeks and Romans, the medieval physician considered the body as a part of the universe, believing that there were four humours, or body fluids, which were associated with the four elements: fire equalling yellow bile; water equalling phlegm; earth equalling black bile; and the air equalling blood. They thought that these four humours had to be kept in balance to ensure good health. Too much of one was believed to cause a change in personality. Too much black bile for example was thought to cause melancholy. Natural functions, such as sneezing, were considered necessary to keep the humors in balance and maintain good health. They believed that a build-up of a particular humour could be corrected through sweating, tears, evacuation of the bowels, or passing urine. One of the reasons why blood-letting was so popular was because it was meant to restore the balance of fluids in the body.

Regardless of their ignorance about the body's functions, their descriptions of diseases were often very good and accurate enough for any modern doctor to give a diagnosis. But recognising the symptoms of an illness does not provide the cure and there was all too often little that physicians could do for their patients; nor could they give much advice about maintaining health or avoiding illness.

Although some doctors were aware that exercise, diet, and a good environment were also important this was a period of history in which people were very superstitious and believed that creatures such as elves and goblins really existed, as well as that the powers of evil were personified by the devil and his demons. In reality many diseases were endemic and were in fact often due to ignorance and poor hygiene, and as the number of people living in medieval towns and cities grew, the situation tended to become worse. Many health problems were simply caused by too many people living together in close proximity in unhygenic conditions and were exacerbated because the underlying causes of infections and disease were so little understood. It was generally believed that diseases were spread by bad odours, whereas the incidences of such diseases as cholera, for example, and attacks of dysentery were frequent due to a lack of clean drinking water being available.

The true causes of infection and contagion were not generally understood at all. Although medical practitioners and both public and religious institutions tried to institute regulations, the lack of any real insight into the true causes of the diseases they were trying to control, meant that few if any effective measures could be taken to protect people's health and the incidences of contagious disease in particular, increased as a consequence. Some of the diseases found in medieval Europe, such as malaria and trachoma, are rarely seen in Europe now.

With some diseases there was an additional factor to be considered. Generally speaking where there are people in large numbers there are also vermin and these can contribute additionally to the transmission of diseases. Furthermore, the lack of a proper sewage system, middens in medieval towns and cities, as well as poor food storage, encouraged fly-borne diseases and all kinds of scavengers – among these were mice and rats. Indeed it is now known that it was the fleas carried by the rats that probably infested every medieval urban conurbation, which led to the pandemic of plague that we now know as the Black Death. This virulent disease, which was in fact plague that manifested

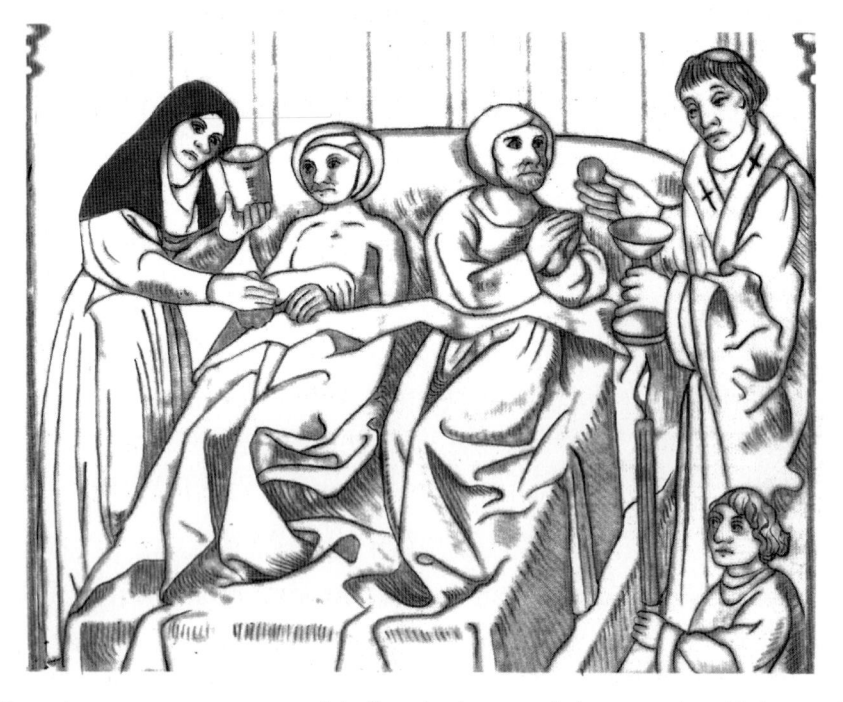

60. Copernicus as a young man was originally only given permission to continue his later studies at university on the proviso that he pursued a course in medicine. Being medically trained, one of Copernicus's duties as a canon would have been to serve as a physician to the cathedral and the local community. There was a strong link between religion and medicine in the Middle Ages. It was often assumed for example that diseases of the body resulted from the sins of the patient – a kind of sickness of the soul. Many people therefore sought relief from their ills through meditation, prayer, pilgrimages, and other non-medical methods. The ministry of the clergy was therefore thought to be every bit as important as that of the physician.

itself in several forms, had first reached Europe during the fourteenth century and small outbreaks continued to recur throughout the Middle Ages.

At the time no-one associated the problem with either rats or fleas and the plague was instead thought to be due to an evil miasma. It was this belief that led to plague-doctors wearing the sinister looking face-masks now associated with the Black Death, that became so common-place during the Middle Ages. This was also the reason why people often carried posies of sweet-smelling flowers, which they held under their noses and frequently sniffed at, in an attempt to avoid breathing in the malodorous vapour of the deadly miasma that they imagined was permeating the air. It is these posies that are being referred to in the children's nursery rhyme, *Ring-a-Ring of Roses*. And the sneezing referred to in the line *atishoo, atishoo we all fall down*, was the first warning sign one received that a person might be infected with the plague.

> *Hear I beseech thee, be favourable to my prayer. Whatsoever herb thou power dost produce, give, I pray, with goodwill to all nations to save them and grant me this medicine. Come to me with thou powers, and howsoever I may use them may they have good success and to whomsoever I may give them. Whatever thou dost grant it may prosper. To thee all things return. Those who rightly receive these herbs from me, do make them whole. Goddess I beseech thee; I pray thee as a suppliant that by thy majesty thou grant this to me.*

61. It is significant that even the Church at times directly propagated pagan practices. Some of these were associated with the origins of the plant medications. The use of medicinal plants was an intrinsic part of an ancient mythical tradition that preceeded the Chrisitian era and there were often special instructions and ceremonies that should be followed when picking certain herbs that were entirely pagan in their origin. There is even evidence of this in the herbals produced by the scribes of both monasteries and convents where incantations not to god but to pagan deities such as a Norse god or an earth goddess occupy the page. Here, in the practice of medieval medicine one can perceive an interesting dichotomy that is repeated in many other aspects of science during those times. It may not be surprising that along with the Church's frequent rejection of the new there is an equally strong adherence to the old: but this did not solely encompass Christian beliefs or values.

Doctors were normally only found in towns and cities or at the courts of various dignitaries. Those living in villages rarely had the opportunity to consult a physician when they were ill. Many treatments therefore were administered by people without a formal medical training. Coroners' rolls from the time recording the various causes of death among the population reveal that lay people often made medical judgments and administered treatments, quite often with disastrous results. But country people were not left entirely without medical help. Herbal remedies, which were commonly used by physicians to treat their patients, probably played an important part in rural medicine too and could be very effective for some ailments.

The use of medicinal plants was an intrinsic part of ancient mythical traditions that preceded the Christian era and there were often special instructions and ceremonies that should be followed when picking certain

62. The intial onslaught of the Black Death killed a third of the population, and even after the worst was over, pockets of the disease remained that caused the plague to periodically reappear throughout Europe for centuries. It killed rich and poor alike, young and old and as this illustration, based upon a contemporary source, clearly depicts, clergy as well as laity. But many people still believed that it was God wreaking his retribution for their sins. Here we see a procession to pray for the end of the plague. In reality supersitious beliefs probably helped contribute to the ignorance surrounding the true nature of the contagion and often to a significant extent hamstrung advances in medical knowledge during the Middle Ages.

63. Herbs had many uses in the Middle Ages. Quite apart from their medicinal properties they were also used to produce love potions and as charms to ward off evil spirits, but also had more mundane but nonetheless useful applications in cooking, as seasoning for some dishes and as preservatives. They were employed to produce scent, provide dyes and as rare pigments for other artistic purposes too.

The value of herbs was well recognised by the Church in the Middle Ages. Most monastery gardens would have had a section devoted to the cultivation of herbs and other useful plants. As many as 250 different species of plants might have been grown in a typical monastery garden or in the precincts of other ecclesiastical establishments. Not all the plants used would have been native species either. Plants from Mediterranean regions and even some from North Africa and the Middle East could be sometimes found in the monastery gardens of Northern Europe: the less hardy plants would have been grown in pots that could be moved inside during the winter. Even opium was sometimes grown to provide relief from pain for those injured or ill.

64. Water crows-foot is just one of thousands of plants that was used by medieval physicians and herbalists to treat the sick. Many of the remedies they employed had their origins far back in time, well before the Christian era. They also remained linked to entirely pagan rituals, that were still being widely applied by those associated with their use – including the clergy.

herbs that were entirely pagan in their origin. This might involve picking a herb at a particular time of day, such as at sunrise, or for example while facing in a particular direction; or there may be a prohibition on looking backwards while collecting some plants. Herbs had many uses in the Middle Ages. Quite apart from their medicinal properties they were also used to produce love potions and as charms to ward off evil spirits, but they also had more mundane but nonetheless useful applications in cooking, as seasoning for some dishes and as preservatives. They were employed to produce scent, provide dyes and as rare pigments or for other artist purposes too.

The value of herbs was well recognised by the Church too in the Middle Ages. Most monastery gardens would have had a section devoted to the cultivation of herbs and other useful plants. As many as 250 different species of plants might have been grown in a typical monastery garden or in the precincts of other ecclesiastical establishments. Not all the plants used would have been native species either. Plants from the Mediterranean region and even some from North Africa and the Middle East could sometimes be found in the monastery gardens of Northern Europe: the less hardy plants would have been grown in pots that could be moved inside during winter. Even opium was sometimes grown to provide relief from pain for those injured or ill.

It is significant that even the Church itself would, at times, directly encourage entirely pagan practices, associated with the origins of the plant medications, which they would advise others to use, or employ themselves.

There is evidence of this in the herbals produced by the scribes of both monasteries and convents where incantations not to God but to pagan deities such as a Norse god or an earth goddess occupy the page. Here, in the practice of medieval medicine one can perceive an interesting dichotomy that is repeated in many other aspects of science during those times. It may not be surprising that along with the Church's frequent rejection of the new there is an equally strong adherence to the old: but this did not solely encompass Christian beliefs or values. Along with the baggage of Christian dogma went even more ancient rituals and superstitions that had preceded the Church's teaching by more than 1,000 years. It is therefore hardly surprisingly that the medieval Church was also willing to continue accepting a model of the universe that even preceded the birth of Christ.

65. In 1513 Copernicus was asked by Pope Leo X to compile a proposal for calendar reform. It had been apparent for some time that the existing calendar could no longer be relied upon. The reason for this was that it had been assumed that there were exactly 365 1/4 days in a year, whereas in reality the true length is slightly less. Thus religious festivals were taking place at the wrong time and this needed to be corrected.

XI

ASTRONOMY

Although Copernicus no longer had the powerful patronage of his uncle the bishop to help him, his special talents were not forgotten by the Church and in 1513 Copernicus was asked to compile a proposal for calendar reform. The Church authorities had been aware for some time that the calendar they had been using could no longer be relied upon. The reason for this was simply due to the fact that it had been assumed that there were exactly 365 1/4 days in a year, whereas in reality the true length is slightly less. Because of this misconception approximately one day had been lost each century since the first century when the Roman Emperor Julius Caesar had implemented his own calendar reform by introducing the idea of adding an extra day every four years originally.

This small error in what is known as the Julian Calendar, by Copernicus's time had resulted in the calendar being approximately ten days in advance of hat it should be. This made it difficult to calculate the correct date and to predict the seasons of the year. Among other things, farmers were in danger of planting their crops at the wrong time and were also unable to decide when harvesting should begin. There were no forecasts or barometers in those days to predict the weather; nor even thermometers by which to measure temperatures. Therefore even a simple matter of anticipating the first frost or the beginning of spring would have been difficult.

The lack of an accurate calendar was considered a particularly serious problem by the Church, which had to rely heavily upon astronomy to help set the dates for all the religious festivals throughout the year. Deciding for

☆ *Polaris*

Direction in which the Earth orbits

North

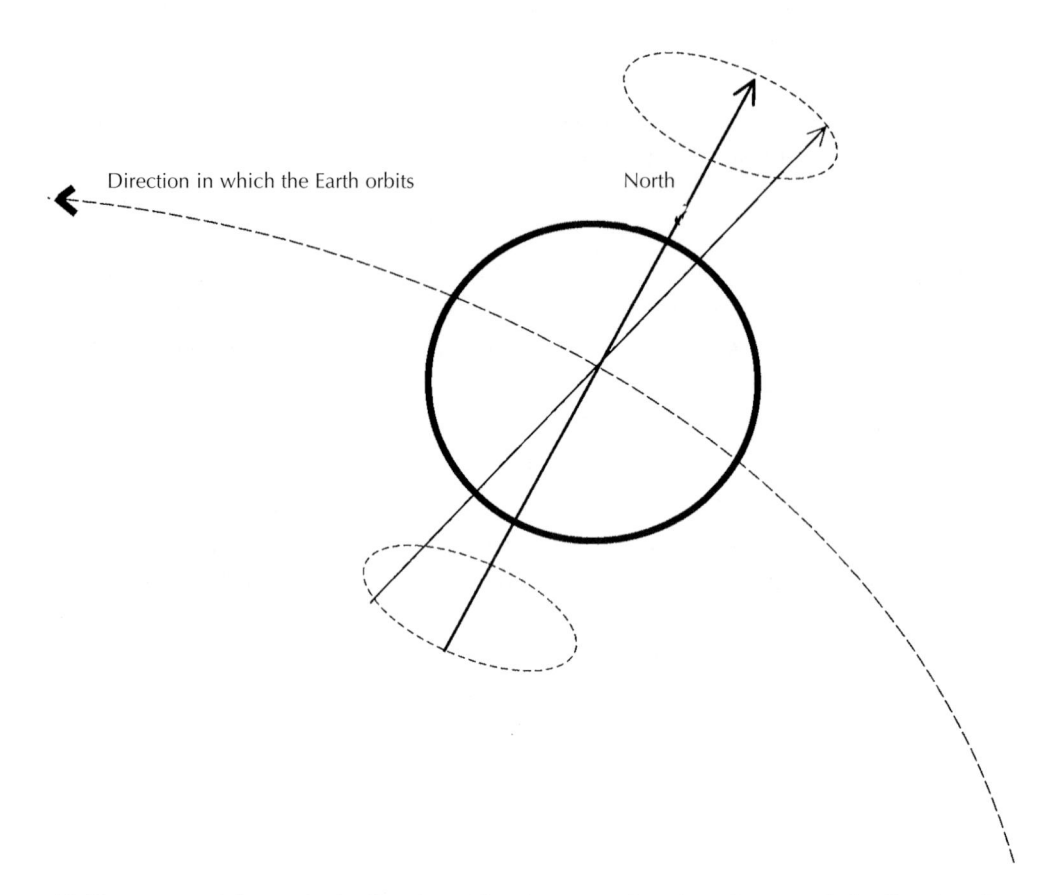

66. The moon and the sun both affect the earth as it turns on its axis and as a result the direction of the axis changes, rather like it does with a spinning top, as it responds to the gravitational pull of the other two bodies. At present *Polaris* (the North Star) is directly above the North Pole but although the angle at which the earth's axis tilts will remain the same, at 23.5 degrees, the direction in which it points will change. This is because the earth's axis gradually shifts, as over a period of 26,000 years, it slowly inscribes an invisible circular path against the background of the stars.

This means that the stars on any date in the calendar will appear in a slightly different position each year, occurring about twenty minutes later than the previous year. As a result the sidereal (or star) year does not synchronise exactly with the solar year. This is most noticeable at each equinox, when the length of the days are nearly the same. This phenomenon is referred to as the precession of the equinoxes.

example when the Easter service should be held, or upon what day Christmas should be celebrated, were major preoccupations.

The link between astronomy and the date of religious festivals may be better understood by explaining the significance and meaning of the term Epact. This is the age of the moon on 21 March, which at that time was considered to be the first day of the year. This, the Church needed to know so that they could to determine when Easter should be. Not knowing when exactly the year had begun therefore created real difficulties. The fact that Copernicus was recognised as a brilliant astronomer and mathematician may have been one reason why he had escaped official criticism from the Church, where many people were already well aware that he had suggested that the earth along with the other planets revolved around the sun and not the sun and planets around the earth, as was the accepted doctrine.

Copernicus had promised Pope Leo that he would try to define the exact length of the year but found that there were difficulties in accomplishing this task. He knew that the true length of a year was slightly shorter than 365 days but he could not work out by exactly what any adjustment should be made. The reason for this was quite simple and was to do with the difference between two separate measurements. One was the solar year, which is the time it takes for the sun to return to a position where it is the same number of degrees above the horizon as it had been previously: for example at midday in midsummer, which is known as the Summer Solstice. This we now know is 365 days, five hours, forty-eight minutes and forty-six seconds. The second is the sidereal year, which is the time it takes for the stars in the night sky to return to the same apparent position as they had occupied before, which is longer by twenty minutes and twenty-four seconds.

The most obvious difference is the variation seen between the sidereal year and the time when the vernal and autumnal equinoxes occur. The vernal equinox is the time each spring when the length of the day and night are nearly equal which normally occurs around 21 March; whereas the autumnal equinox is when the same phenomenon occurs on about 21 September. Because the sidereal year is slightly longer than the solar year the equinoxes, in relation to the position of the stars, are progressively later by the equivalent of twenty minutes and twenty-four seconds every successive year. In other words the position of the stars is the same as the one they were seen to have occupied twenty minutes and twenty-four seconds earlier the previous year. This may not seem a lot but over a period of several centuries this becomes quite noticeable.

This discrepancy in the solar and sidereal year is due to the fact that the earth's axis is slowly swinging around in response to the combined gravitational

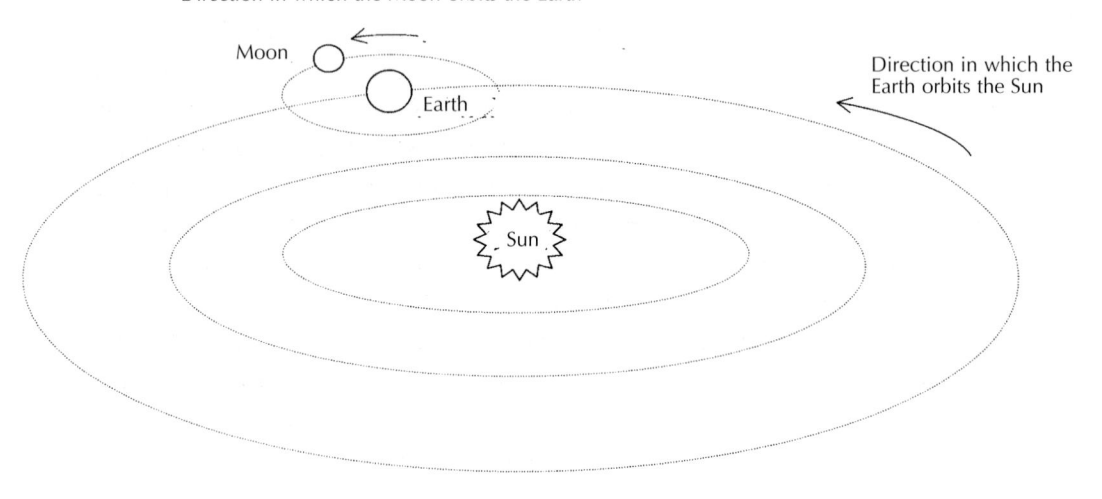

67. While the moon is orbiting the earth, the earth in its turn is orbiting the sun. Therefore by the time the moon has completed its journey around the earth the earth itself has moved on. This means that the moon as it orbits the earth must travel slightly more than a full circle before it completes its cycle of phrases. For this reason a complete orbit of the earth does not take the moon back to exactly the same position as it occupied at that point previously against the background of the night sky, because the movement of the earth orbiting the sun slightly alters the position in which the stars appear. This discrepancy can be detected when comparing the position of the moon against the background of the stars and is referred to as the lunar precession. In addition, as a result of this, the stars also appear about four minutes earlier each successive night.

pull of the moon and sun. This change had been first recorded in the second century BC by Hipparchus of Alexandria, who noticed that the earlier observations made by Babylonian astronomers living 190 years before did not tally with his own. He realised that the direction of the earth's axis slowly changes over a period of 26,000 years. One effect is that the position of the Pole Star, which also known as *Polaris*, does not remain the same. In about 12,000 years for example the north celestial pole will be pointing approximately in the direction of a star called Vega and 26,000 years from now it will again point to *Polaris*. The results of this gradual shifting of the earth's axis are referred to as precession and the variation between the time of the equinoxes, in relationship to the position of the star, is known as the precession of the equinoxes.

Finding the precise length of the year was therefore a problem because the seasons by which we mark the solar year are determined by the amount the earth's axis tilts each hemisphere towards or away from the sun at different points during its orbit; whereas the apparent position of the stars, by which the sidereal year is measured, depends upon the time it takes for the earth to complete an orbit around the sun. The precession caused by the small changes in the orientation of the earth's axis that occur within this period means that these two do not appear as being in exact synchronisation. This must have been especially frustrating for earlier astronomers as the movement of the stars across the heavens was the most accurate timekeeper they had before the invention of the first modern timepieces by Harrison in the eighteenth century. It is also worth noting that, although we now know what causes these differences, Copernicus could not have known, as at this time he still had not even obtained the absolute proof he wanted to demonstrate that it was the earth that really revolved around the sun and not the other way around.

Copernicus, as with other astronomers of his time, knew that the movements of heavenly bodies did not always conform to the precise expectations one might have – and the variations seen between the timing of different annual events was already well documented and allowed for in any astronomical calculations, or observations. What was not always understood was why they occurred or how they fitted into the greater scheme of things. This meant it was difficult for Copernicus to decide by which yardstick the year's length should be measured.

There is for example a similar discrepancy between the phases of the moon by which we measure the lunar month and sidereal month: which is the time it takes for the moon to return to the same apparent position when seen against the stars in the night sky. This is because the earth is rotating as it orbits the sun, which means that the moon as it orbits the earth must travel slightly more than a full circle before it completes its cycle of phases. Even the solar day and the sidereal day differ in length for a similar reason – because the rotation of the earth orbiting the sun slightly alters the position in which the stars appear – and as a result of this the stars appear about four minutes earlier each successive night. The fact that the vast majority of people in the medieval world did not realise that the earth orbited the sun meant they had no way of understanding why the lunar month and the sidereal month were different; nor why the sidereal and solar day were different lengths either.

What might have also been preventing Copernicus from resolving some of the problems with his own theory was his belief that a universe, which he accepted as being created by God, should have any imperfections or aberrations in its design or functions. This assumption, as we shall see, was to later cause him further difficulties with his own theory of a sun-centred cosmos and could have been one reason why he delayed its publication.

One very important influence in medieval astronomy was Regiomuntanus whose writings Copernicus would have referred to when a student. Copernicus also referred to the astronomical tables compiled by Regiomuntanus to help construct his heliocentric model. Although remembered as an astronomer it would appear that Copernicus did not make that many observations of his own during his lifetime; or at least not until after his return to Frauenburg.

Copernicus seems to have based the vast majority of his calculations upon the work of earlier astronomers and drew his conclusions about how the stars, sun, moon, earth and the other planets each move, as a result of checking and reworking their figures. This even included studying the records of Ptolemy whose theories he was trying to disprove, as well as probably those made by the early Greek astronomers such as Archimedes and Aristarchus. We know from Archimedes' own writings that these two men had sat together in the market place in Alexandria and discussed the possibility that the earth and the other plants orbited the sun, to form what is called a heliocentric system, many centuries before Copernicus resurrected the idea. They had not, though, been able to offer any proof of this: it was merely a hypothesis.

What Copernicus was trying to do was use his skill as a mathematician to provide the proof required to finally confirm the heliocentric hypothesis. In fact in an attempt to achieve this Copernicus even used the observations of Ptolemy as a source of raw astronomical data even though he disagreed with his conclusions. We also know that he was aware that Aristarchus had suggested that the earth orbited the sun sixteen centuries earlier, because Copernicus made a note of this in the margin of one of his early manuscripts. What we do not know, is whether or not Copernicus first conceived the notion of a heliocentric system entirely by himself, or got the idea originally from the ancient Greek manuscripts he had studied as a student.

68. Leonardo da Vinci was not only an outstanding artist; he was also an architect, an engineer and a scientific genius. His sketchbooks show designs for a vast range of inventions including ideas for building battle tanks and even flying machines. One drawing was a design for a helicopter. Leonardo da Vinci also made one interesting contribution to astronomy. It had been noticed that when there is a crescent moon the rest of the lunar surface often seems to be illuminated with a faint, diffuse light. Leonardo correctly deduced that this was earthlight.

XII

CONTEMPORARY INFLUENCES

One should not forget that Copernicus was not the only innovative thinker of his time. There were others equally influential in both the arts and sciences. The celebrated artist Leonardo da Vinci was for example one of Copernicus's contemporaries. He is best known for the *Mona Lisa*, which is possibly the most famous painting in the world. Another well-known painting of his, *The Last Supper*, is a panorama showing Jesus and his disciples sharing their last meal together before Jesus was arrested and taken for crucifixion. A total of seventeen paintings by Leonardo da Vinci still survive as either copies or originals to the present day.

Moreover, Leonardo da Vinci was not only an outstanding artist he was also an architect, an engineer and a scientific genius. His sketchbooks show designs for a vast range of inventions including ideas for building battle tanks powered by horses and even flying machines. One drawing was a design for a helicopter. These sketchbooks are now owned by the Queen of England. There is no record of anyone trying to construct what were then such futuristic objects: the kind of materials and technology required were simply not available in the fifteenth century when da Vinci lived. Copernicus was not therefore by any means the only original thinker of his day, nor the only one to be multi-skilled.

In fact, although not an astronomer, Leonardo da Vinci did make one interesting contribution to this subject. It had been noticed that when there is a crescent moon the rest of the lunar surface often seems to be illuminated with a faint, diffuse light. Leonardo correctly deduced that this was earthlight: where the earth was reflecting the rays of the sun onto the face of the moon.

Probably one of the most important influences to emerge during the latter part of the Middle Ages, when Copernicus lived, was the Humanist movement. For the Humanists of the Middle Ages the knowledge represented in the wisdom of the ancient cultures of Greece and Rome, found in the copies of manuscripts from these past civilisations, provided an authoritative viewpoint outside of that previously represented, solely by the Roman Catholic Church. Moreover, it also encouraged them to seek solutions and enlightenments through dint of human endeavour rather than relying solely upon spiritual experience or religious dogma. Indeed, the Humanists were concerned with the development of what they perceived to be the human virtues of all forms. The promotion of education, knowledge and understanding were part of their philosophy but so were such human characteristics as fortitude, honour and judgement and many more.

Not withstanding the fact that many Humanists were clerics too, the ideas provided by the rediscovery of texts from the ancient civilisations of the ancient world, offered real alternatives to the often hidebound thinking taught by the Church, as well as the added stimulus provided by contact with the Islamic world. Familiarity with other ideologies and the wisdom of bygone cultures encouraged many intelligent people to rely far more upon their own powers of reasoning than had been hitherto normal – and be at times as inquisitive as those early natural philosophers of the ancient world and ask equally searching questions about the universe and the world in which they lived. The overall value of Humanism as a movement was its recognition of human potential and a commitment to encouraging its development rather than prescribing its limitations, which had a big part to play in creating a more positive attitude towards the acceptance of new ideas and learning.

The other great influence, destined to reshape the religious map of Europe was the Lutherans, whose movement heralded the reformation that was eventually to create a schism in the Christian Church. Martin Luther like all other clergy was originally a pastor of the Roman Catholic Church. The Church though – and in particular the Vatican in Rome – had become increasingly corrupt over the centuries, with power and wealth being more important to some clergy than godliness. This even included some popes. There were those who realised this, but in sixteenth-century Europe it was the only recognised Church and its teachings and authority as a religious institution were rarely seriously challenged. Nevertheless, in 1517 that is exactly what Martin Luther did. He nailed a proclamation to the door of his church at Wittenberg in Germany.

69. Clergymen, even popes, were not always completely altruistic. Here we see how they abused their position by selling forgiveness to the public – this practice was known as the selling of Indulgences. This made the Church a vast amount of money and at the same time brought it an equal amount of turmoil.

The proclamation had ninety-five separate sections but the most important part dealt with what were called indulgences. Some members of the Roman Catholic Church, including the popes, were making money by selling documents known as indulgences that promised people that they would be forgiven for their sins and avoid being sent to hell when they died. The pope was even claiming that he could arrange forgiveness for those who had already died, providing that enough was paid as an inducement by their relatives, by reducing the time spent by those languishing in Purgatory – which is the place where people were meant to purge, or get rid of their bad habits and wicked ways.

Everyone does things that are wrong sometime in their life and most people in the Middle Ages really believed in Heaven, Hell and Purgatory and thought that they would have to pay for any wrongdoing in the afterlife; even though as Catholics most went regularly to confession to seek God's forgiveness for their sins. Furthermore, the vast majority of ordinary people had been brought up to accept what the Church said without question and now some unscrupulous and greedy clergy in the Roman Catholic Church were playing on their fear to make money. In modern-day terms we would call this a scam! And that is basically what Martin Luther called it.

Many people supported what Martin Luther said and did, including the Humanists. He even had help from one of the German princes who hid him

70. Martin Luther was originally a pastor of the Roman Catholic Church. The Church though had become increasingly corrupt over the centuries. Nevertheless, its authority as a religious institution was rarely challenged but in 1517 Martin Luther did just that when he nailed a proclamation to the door of his church at Wittenberg in Germany. It contained ninety-five separate sections and the most important part dealt with the practice by some clergy of selling indulgences, which claimed to offer people forgiveness for their sins.

in one of his castles to prevent anyone trying to harm him. But the official Church of Rome was not pleased. After failing to get Luther to recant his beliefs, the pope in Rome sent him an official document known as a papal bull excommunicating him. This meant that he was being expelled from the Church; not only as a priest but also as a Christian. Most people in medieval Europe would have been terrified if this had happened to them, because they believed that this would condemn them to eternal damnation and the torments of everlasting Hell when they died. Martin Luther simply tore it up and threw it away!

The result of all this was that Luther and his many supporters broke away from the Catholic faith to establish a new Reformed Church. Eventually because of their protests, the Lutherans and others who disagreed with the practises of the Roman Catholic Church became known as Protestants and in Europe entire regions turned their back on the Church in Rome and became Protestant. Ultimately warfare would result between Catholics and Protestants and entire countries would take sides and be drawn in to the conflict, with this on occasions leading to civil war. In 1527 there was even an attack by unpaid mercenaries on Rome itself in which it was said that many Lutherans were involved. This resulted in wholesale rape and torture and thousands of innocent Roman citizens died. But in the short-term, although some

Catholic clerics were outraged by what they saw as heresy, others tried to build bridges and take a more tolerant attitude towards those whose interpretation of the Christian religion differed from their own. This in some quarters actually encouraged a more lenient and open attitude towards people like Copernicus who it was recognised simply sought a new knowledge and understanding of the world and the universe in which they lived. This though was not always the case and in places the persecution of other Christian dissenters continued under various guises with the Spanish Inquisition being the worst-case scenario.

Strangely enough the rebel priest Martin Luther, although he himself had broken away from the Catholic Church, was equally critical of Copernicus and his theories. The fact that Copernicus had suggested that the all planets, including the earth, revolved around the sun and not the sun and other planets around the earth, Luther thought was completely ridiculous. To him and many other people it seemed obvious that the sun went around the earth simply because one could observe it passing across the sky each day; the fact that a revolving earth would create the same effect was discounted: one could not detect any evidence of the earth moving. The problem was that many people believed that one would be able to sense some kind of movement just as you would when riding a horse or travelling in a wagon and for example, feel the wind in your hair.

Luther's rebellion may have led to him rejecting the authority of the Church of Rome on moral grounds, as a direct result of what he saw as corruption; whereas Copernicus's rebellion was an intellectual one. He was not rejecting the authority of the Church but he was questioning one aspect of its doctrine – that God had created a universe with the earth at its centre. Luther objected to this as well; but as we shall see later not all Protestants were in accord with Luther's criticisms of Copernicus and his new ideas about earth's true place in the universe.

71. The fortifications at Frauenburg cathedral were built because of the city's long history of conflict involving the Teutonic Knights. Copernicus lodged in one of these turrets and the battlements provided him with a good viewpoint from which to study the night sky.

XIII

WARFARE

Frauenburg, where Copernicus lived out the remainder of his life, was in the duchy of Ermland where his uncle had once ruled as bishop. Officially part of Poland, Ermland was actually one of four duchies, two thirds of which were still largely controlled by the Teutonic Knights who were meant to show allegiance to the Polish crown, but this had never sat easily with the knights.

The reader may recall that in 1225, long before Copernicus was born, one of the Polish princes, Konrad of Masovia, who had been having trouble with incursions of the heathen Prussians into his territories, had invited the Teutonic Knights to subdue the Prussian tribes, or Prusi as they were then known, and settle the lands of the people they conquered as his subjects. As an order of warrior monks dedicated to eradicating all 'Enemies of Christendom' the knights had accepted the prince's invitation. But it had not been easy and after the hard fought war had been won and the native Prussians subdued, once the Teutonic Knights had taken possession of their lands they had not been so keen to recognise the sovereignty of the prince or of later Polish monarchs over the regions they had won in battle.

The knights were originally hospitallers, which meant that they automatically received absolution for their sins from the reigning pope, Innocent III. Although this privilege was granted way back in 1211, it continued to give them a unique political advantage centuries later. It made it very difficult for any prince or bishop to exercise control over their excesses or demand their

72. In 1211, Pope Innocent III granted papal exemption to a then newly formed order of hospitallers in the Holy Land, known as the Teutonic Knights. This meant that they only had to answer to the authority of the incumbent pope. This exemption, still in force at the time of Copernicus, was to be a major factor influencing the politics of Ermland over which his uncle ruled as bishop.

Pope Innocent III was also responsible for launching two crusades. One was against Muslim forces that were threatening Christians in the Near East, but a later crusade in 1209 was against fellow Christians and intended to suppress the Albigensian heretics in southern France, who denied the authority of the ecclesiastical hierarchy. This instigated a new chapter in the history of the Church when he as Pope first sanctioned the repression of heresy by force.

This ecclesiastical approach to dissent was to culminate later in the formation of the Inquisition. The climate of fear thus created was to make the majority of people far less inclined to question the received wisdom of the Church. There is moreover little doubt that it caused some of the more innovative thinkers to hesitate before broadcasting their ideas or even sharing their thinking with others, which most probably adversely affected developments in medicine, science and many aspects of technology throughout the medieval period.

allegiance; nor could they legitimately impose any sanctions on individuals, or the order itself, without resorting to force of arms.

One problem was that in the course of the war and its brutal and vengeful aftermath, the vast majority of the Prussian population had been killed – men, women and children had been slaughtered by the noble Christian Knights. Prussia, by then desolate and largely depopulated, was quickly resettled by colonists invited in by the Teutonic Order. Most of these were from Germany, where the knights themselves originated. This had increased the number of native German-speakers occupying these territories, giving them a strong power base, which further encouraged the knights to try and resist repeated attempts to impose Polish sovereignty. Many battles had been fought over the disputed lands and the matter had only been resolved when the knights had been defeated, as mentioned earlier, at the battle of Tannenberg in 1410 by the joint forces of the Polish-Lithuanian alliance.

The hill on which the cathedral was built had already been fortified by the Teutonic Knights after they had originally conquered the native Prussians, who they from then on held in oppressive serfdom. Over the years though there had been a succession of other conflicts that had arisen between the Teutonic Order and their nominal overlords – the Polish monarchy. The town was strategically quite important and had suffered several times during these skirmishes. Eventually, for this reason, a massive wall was constructed to protect the cathedral from further attack and turrets were set in the wall to aid in its defence. Copernicus had accommodation in one of these turrets when he finally returned to take up his post as a canon and it was from this turret vantage point that he was to make many of his later astronomical observations but before long Copernicus was to be rudely reminded of the true purpose of those fortifications.

One of the first things Copernicus had to do when he returned to Frauenburg was to swear allegiance along with the rest of the canons of the cathedral to the new king of Poland who had just recently succeeded to the throne. The duchy of Ermland though was a troubled land and change was in the air in more ways than one. Discontent had been festering for generations in Ermland with sporadic outbreaks involving armed conflicts and rebellion on the behalf of various parties. Diffused with political intrigue the land was like a powder-keg awaiting the touch-paper.

The Teutonic Knights still wanted greater independence from Poland and those they held under their sway wanted freedom from the oppression they had suffered from the order themselves. In 1519 warfare erupted once again. Albert, the grandmaster of the Teutonic Order, had been secretly recruiting mercenaries whom he had sent as raiding parties into Ermland to loot and pillage the defenceless population. The Polish king, Sigismund, responded by sending in his own troops in defence of the duchy.

73. Allenstein Castle was one of the Bishop of Ermland's residences. Copernicus found himself faced with the responsibility of defending the building when it was being attacked by the Teutonic Knights, while they were waging war against Poland. The knights had laid siege to the castle but although Copernicus had only two other clerics and a few servants to help him man the defences, the siege was not sustained. This may have been because it was discovered that the defenders had acquired firearms. These were never used, because there were no bullets available; but the soldiers attacking need never have known this.

74. When the grandmaster of the Teutonic Order began recruiting mercenaries and sending them as raiding parties into Ermland to loot and pillage the defenceless population, the Polish king, Sigismund, responded by sending in his own troops in defence of the duchy. Full-scale warfare ensued and the peasantry of Ermland were the losers. Their country became a battleground with both sides slaughtering anyone whom they suspected of siding with their enemies. The two armies rampaged through the countryside, burning farms and destroying crops. The marauding soldiers made certain that they took anything of value first, before killing any peasants who fell into their hands and raping the women. They slaughtered their livestock or otherwise stole them along with everything else.

It was only through the mediation of the Holy Roman Emperor that an armistice was eventually declared in 1521. In the aftermath of the war there was much to be done and Nicholas Copernicus was soon involved in the efforts that were then required to repair the damage done by the war and re-establish order in the land. In recognition of his efforts he was awarded the title of Commissar of Ermland and set about getting the country back on its feet. Part of his job was supervising the repatriation of those who had fled the war, seeing that they were re-housed and supervising the rebuilding of homes that had been destroyed. Food shortages, resulting from the abandonment of farms, the loss of livestock and the slaughter of peasants and their families, also had to be remedied if the population was not to starve.

Copernicus was also given the task of listing all the complaints the population of Ermland had against the Teutonic Knights. These were considerable as the knights had destroyed towns, burnt villages and homesteads, stolen food and animals, robbed and pillaged and had also murdered many innocent people – very often whole families had perished at the hands of these warrior-monks and their troops.

At this time Tartars were causing the king problems too with incursions across the Russian border. Once this issue was resolved, King Sigismund sent for the grand-master who was still technically his vassal, demanding a meeting at Torun where Albert should pay him homage. But when Albert failed to appear the outraged king sent Polish troops to invade the territory of the Teutonic Order and full-scale warfare ensued.

The peasantry of Ermland were the losers. Their country became a battleground with both sides slaughtering anyone whom they suspected of siding with their enemies. Both armies rampaged through the countryside, burning farms and destroying crops. The marauding soldiers made certain that they took anything of value first, before killing any peasants who fell into their hands and raping the women. They slaughtered their livestock or otherwise stole them along with everything else.

Most of the canons at Frauenburg Cathedral had sought refuge elsewhere when hostilities began but Copernicus and one other ageing canon stayed on. While the war raged all around him Copernicus continued his astronomical studies of the stars and planets. Relatively few fortified towns were attacked, the Teutonic Knights and their protagonists preferred easier targets, such as isolated hamlets and unarmed peasantry. But there were exceptions and before long Frauenburg itself was under siege. Nevertheless, it is said that even then Copernicus was making observations while the battle for the town was being fought beneath the very walls of the cathedral and continued to do so even when the town, which had been set on fire by the victorious knights, was ablaze.

Shortly after this event, though, the war began to go badly for the Teutonic Order. The Tartars had been repulsed and peace had been made by King Sigismund with Russia which meant that more Polish troops were then available to fight the knights and as a consequence Albert, the grandmaster, began to sue for peace. This lull did not last however, as in the meantime Albert had learnt that additional help was on its way from Germany and as a consequence he decided to recommence hostilities. By now Copernicus's duties had taken him to Allenstein Castle and this time he really found himself in the thick of the fighting when the grandmaster decided to lay siege. Although there were only two other canons and a few servants with Copernicus this small group of defenders, nevertheless, managed to successfully resist all attempts to take the castle. They even had the very latest armaments of the day – crude firearms. These had been smuggled into the castle by a secret sympathiser; but without bullets. These were meant to come later but in the event they were never needed, as fortunately the knights' efforts to besiege the castle were not maintained.

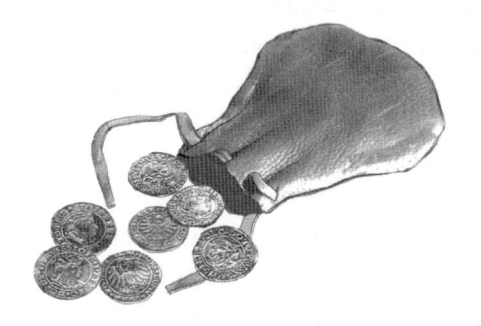

75. Among his other talents Copernicus was an economist and wrote two treatises on minting money, one in 1522 and another in 1528 when there was a problem with counterfeit coinage that had come into circulation following the war in Ermland.

It was only through the mediation of the Holy Roman Emperor that an armistice was eventually declared in 1521. By this time all parties were heartily sick of fighting and peace was finally established in Ermland. To mollify him, the grand-master of the Teutonic Knights was made a duke and in return swore allegiance to the Polish sovereign and Prussia became a fiefdom of Poland. But in the aftermath of the war there was much to be done and Nicholas Copernicus was soon involved in the efforts that were then required to repair the damage done by the war and re-establish order in the land. He was awarded the title of Commissar of Ermland in recognition of his efforts and set about getting the country back on its feet. Part of his job was supervising the repatriation of those who had fled the war. Such people had to be re-housed on their return and homes that had been destroyed had to be rebuilt. Any food shortages, resulting from the abandonment of farms, the loss of livestock and the slaughter of peasants and their families, also had to be remedied if the population was not to starve. In fact the situation then in Ermland was not unlike that that had faced the authorities in the Balkans after the recent disastrous war that destroyed what was formerly the country of Yugoslavia.

Copernicus was also given the task of listing all the complaints the population of Ermland had against the Teutonic Knights. These were considerable as the knights had destroyed towns, burnt villages and homesteads, stolen food and animals, robbed and pillaged and had also murdered many innocent people – very often whole families had perished at the hands of these warrior-monks and their troops. This was an important task as the authorities had every intention of demanding restitution from the Teutonic Order to compensate those they had wronged during the war.

He was also from time to time given other responsibilities. Copernicus was still highly respected elsewhere and often in the highest political and social circles. Both kings and popes sought his advice. In 1526, following the cessation of hostilities, Copernicus was asked to assist the king's secretary with the mapping of the kingdom of Poland and the duchy of Lithuania. This is not surprising because Copernicus was a very capable mathematician who the authorities knew could be relied upon to make accurate measurements. He was also as an astronomer familiar with a particularly useful branch of mathematics known as trigonometry.

Trigonometry is a branch of mathematics employing geometry to calculate among other things angles and distances, relying on the properties of triangles. Copernicus would have used trigonometry for example to compare the apparent angular distances between stars or to measure the height of celestial bodies above the horizon and to track the movement of stars and planets across the sky. It is called trigonometry because it uses the properties of triangles (in Greek, *trigonon*) to make such calculations. One of the simplest of scientific instruments of the time, the cross-staff actually used a very basic form of trigonometry both to take astronomical measurements and for surveying. The same methods employed in astronomy can be used to measure angles and calculate the distance between terrestrial objects too, or find the height for example of a hill or mountain. Copernicus would probably have also used another instrument that he was very familiar with. This was an astrolabe, which apart from its value to astronomers and mariners could be employed by surveyors and cartographers to measure distance: once more utilising the principles of trigonometry.

Copernicus seems to have abandoned his astronomy at certain times during his life, but was never idle. Among all his other talents he was also something of an economist. He had already written a paper on minting money in 1522 and in 1528 he presented a further treatise on coinage. After any armed conflict chaos tends to reign. It was no different with the war in Ermland. One of the problems was with counterfeit or poorly minted money. Copernicus, whose family had been involved in commerce in Kraków was very familiar with money matters and he realised that this could seriously harm the country's economy and delay its recovery after the war. Copernicus therefore wrote the two treatises on coinage and the minting of money warning of the dangers. Unfortunately his ideas were never acted upon but it is interesting to note a particular comment he made that, 'bad money drives out good'. This is known as Gresham's Law by economists, because it was also said much later by one of Elizabeth I's courtiers, Sir Thomas Gresham; but Copernicus had said it first – thirty years earlier.

76. One of the people who had helped Copernicus with his duties when he served as Commissioner of Ermland was another canon, called Tiedemann Giese. He was a gentle man and proved to be a good friend of Copernicus throughout the astronomer's life, being one of his most staunch and powerful supporters. Destined eventually to rise to the position of bishop, he became very influential in the Church. Indeed, his long-time friend the prelate was one of those who eventually persuaded Copernicus, despite his fears of ridicule and persecution, to publish his theories of a heliocentric planetary system.

When the Reformation began and Martin Luther threatened to divide the Church and set Christian against Christian, Canon Tiedemann Giese was one of those who made a public plea for tolerance and reconciliation in a small book that he had published, that began with the words 'I reject the battle'.

XIV

FRIENDS AND FRATERNITY

One of the people who had helped him earlier with his duties of Commisionar of Ermland was another canon, called Tiedemann Giese. He had become a valued friend and in later years was to play an important part in Copernicus's life. He was also to become very influential in the Church. Tiedemann Giese eventually rose to the position of bishop and remaining a good friend, he became one of Copernicus's most staunch and powerful supporters.

In 1523 his uncle's successor, Bishop Fabian von Lossainen, died and was succeeded by Bishop Ferber. Bishop von Lossainen had been a rather tolerant man who was not unduly concerned about Copernicus and the new ideas he and his associates were promoting. Whereas Bishop Lossainen's attitude towards Luther was to merely consider him a learned monk with his own opinions of the scriptures, Bishop Ferber's approach to any such form of dissent was very different. He strongly opposed Luther and his reformist policies: which he saw as undermining the previously unchallenged authority of the Church in Rome. His dislike, moreover, extended to anyone who questioned the established teachings of Catholicism – and that included Copernicus and his supporters. His first edict as bishop, issued in 1524, for example stated that anyone who paid any attention to those who might divide the Church or questioned its absolute authority would be condemned to eternal damnation.

Not everyone in the chapter at Frauenburg felt the same way though, Canon Tiedemann Giese was one of those who fervently disagreed with this approach and later made a public plea for tolerance and reconciliation in a

77. Copernicus served under several bishops after his uncle Bishop Lucas Waczenrode died and one of these was Bishop Danticus. Although he had been quite a libertine as a young man and kept a mistress, he forced Copernicus to send away his housekeeper Anne Schilling. She was a niece of Copernicus and the relationship could have been perfectly innocent but Bishop Danticus may have been more concerned with appearances than actualities.

small book that he had published, that began with the words, 'I reject the battle'. Tiedemann Giese was a gentle man and proved to be a good friend of Copernicus throughout the astronomer's life. He offered him moral support with his work, involving his new model of the cosmos, but also when in 1538 Bishop Ferber died, a new Bishop Lucas von Danticus, had taken his place. Following this appointment the good-hearted Tiedemann was soon to be called upon to intercede for his friend Nicholas in a far more delicate matter.

It has already been mentioned that many of the clergy in medieval Europe disregarded their vows of celibacy. The new bishop of Ermland had, himself, taken a mistress in his younger days and was the father of two children. Nor did he make any secret of this as he regularly sent money to his ex-mistress so ensure that the needs of herself and her children were catered for. He had also had their portraits painted so that he could better remember his little family. Regardless of his obvious fondness for his offspring and their mother, for Bishop Danticus such temptations were a thing of the past, more understandable and excusable in a young man new to the priesthood maybe than in a figure of authority who had to set an example. He was therefore dismayed to find that two of his canons, both older men, had apparently strayed – and one of these was Nicholas Copernicus.

The other errant canon was Alexander Sculteti, who actually lived quite openly, as if he were a married man, with his mistress and the several children who were the result of their liaison. Such a flagrant disregard for public morality by a canon of the Church could not be ignored, even in those sometimes inconsistent, but nonetheless surprisingly permissive, times. Moreover, the two canons Sculteti and Copernicus were known to be close associates, with the two men sometimes visiting one another's dwellings. Therefore, Nicholas Copernicus was probably to some extent already condemned by association in the eyes of Bishop Danticus. Consequently, when the new bishop discovered that the ageing Canon Copernicus also had a woman living with him, he naturally assumed the worst: that Copernicus too was keeping a mistress.

Copernicus claimed that she was merely his housekeeper. He was already sixty-three years old and so by this time it was quite possible that this was no less than the truth. The woman's name was Anne Schilling and she was, moreover, a distant cousin of the canon. None of this though mattered to the bishop who was probably more concerned with appearances than actualities and so Copernicus was instructed to send her away. Bishop Danticus demanded that Canon Sculteti's long-time mistress left too along with her children. Naturally enough, neither woman was happy about this and it became obvious that neither was going to leave willingly, nor without complaint.

Copernicus, true to his character, did what he always did when faced with a dilemma, played for time and procrastinated. A series of letters passed back and forth between him and his bishop without any action being taken. Eventually Tiedemann Giese was asked by an exasperated Bishop Danticus to intercede. By now Tiedemann Giese was himself a bishop at Kulm and without mentioning the communication he had had with Bishop Danticus, he wrote to Copernicus to try and persuade him to comply with his bishop's instruction. Canon Sculteti, as Tiedemann Giese pointed out, was already engaged in another conflict with Bishop Danticus regarding the chapter's feality to the Polish crown. This Sculteti opposed and Tiedemann warned his friend against the dangers of being drawn into that greater dispute too. It proved to be wise counsel as this rebellion by Sculteti was to result not long after in the whole chapter being temporarily excommunicated. In fact Canon Alexander Sculteti was to be later expelled from the chapter following these events. Copernicus did eventually reluctantly accept Tiedemann Giese's advice and Anne Schilling moved out; but even then did not leave Frauenburg until six months later and only after there had been further rumours and complaints about she and Copernicus being seen meeting and consorting together.

Two very typical characteristics of Nicholas Copernicus's nature are made evident in this little saga – his indecisiveness and his stubbornness – features of his personality that were reflected also in the guarded attitude he displayed concerning his ideas in astronomy. It shows too that, although by nature quite timid and normally subservient to authority, he could also be both rebellious and secretive when it suited him: exactly as he was as an astronomer, sticking to his own theories about the earth's place in the universe, even in the face of opposition, but uncommunicative about his research and too constrained by his own uncertainties to broadcast his thoughts to a wider audience.

78. The triquetrum could be used to measure the apparent distance between different celestial objects, as well as giving their height in degrees above the horizon. This was the same kind of instrument that had been used centuries before by Ptolemy, the Egyptian astronomer, and the construction of the triquetrum was described by him in his manuscripts which were still being referred to by astronomers during the Middle Ages. For this reason the triquetrum was sometimes known by another name – Ptolemy's rulers. It consisted of three straight bars with markings which could be reconfigured in order to take the measurements required.

XV

COPERNICUS THE ASTRONOMER

Copernicus had no telescope. This instrument would not appear for nearly another 100 years. For most of his observations he used the same kind of apparatus that had been used in ancient Greece many centuries before by the very astronomers whose writings he had studied as a student and whose work he continued to refer to when developing his own ideas about the cosmos. Copernicus would have also used the astronomical instruments of his day such as the astrolabe and the celestial globe and would have presumably have had too a nocturnal to tell the time during his night-time observations. Although earlier European astronomers had sometime used a device known as a cross-staff to make their measurements, we know that Copernicus relied to a large extent upon an instrument known as a triquetrum, for many of his observations, which was far more sophisticated.

He had probably first seen an example of this instrument while a student at Kraków University when, as mentioned earlier, the rector had received a triquetrum among several other astronomical instruments sent as gifts by a successful former student. Copernicus actually had to build his own from the description given by Ptolemy in his treatise on astronomy. The reader may recall that the other name given to the triquetrum was Ptolemy's rulers.

The triquetrum had an advantage over the cross-staff for two main reasons. Firstly it was permanently fixed in one position, which meant that any observations could be more easily compared with one another because there would be no variation in the position from which two or more observations were made. Moreover, unlike the cross-staff, which was hand-held, one could be certain that with a fixed piece of equipment like the triquetrum, the angle recorded when it was being used would remain absolutely the same for all observations and measurements taken.

In fact, Copernicus did not make very many astronomical observations during his life and seemed to actively dislike doing so. It is said that he had poor eyesight, which certainly may have been so towards the end of his life. Even if he had taken to wearing spectacles they could not have improved his view of the stars and planets. The cloudy weather often experienced in Northern Europe could not have helped either. Frauenburg in particular, being situated by a fairly large fresh-water lake, would most likely have been subject to frequent mists too.

There was, moreover, little encouragement. Few people agreed with his views or even understood his reasoning. Even among those friends and acquaintances who offered him support there were not many with whom he could really share his ideas and interest in astronomy and few to whom he could communicate his thoughts as a fellow astronomer, or engage in professional discourse. His good friend Tiedemann Giese may have been the one exception as it is known from a letter written much later by a friend that he not only possessed an armillary sphere but had also had gone to the trouble of purchasing a particularly fine sundial from England.

Elsewhere, though, he did have his fans among those few well-informed people who knew of his revolutionary theory. Although he had not yet officially published his theory, of course he had raised the idea in a paper he had written as a young man for some of his colleagues and friends while at Bologna University and had also circulated an early version of his theories in the *Little Commentary* he had written when living with his uncle the bishop at Heilsberg Castle. We know too that news had also leaked out about the new version of his theory he was working on while at Frauenburg because in 1536 he received a letter from one of his admirers – and from a somewhat surprising source.

The letter was from Cardinal Schonberg in Rome and was very complimentary about the work Copernicus was doing as an astronomer. He also asked Copernicus to send him copies of his manuscripts on astronomy and promised to pay for any expenses incurred. Copernicus was very proud of this letter and it was included in the book he finally produced outlining his theories. It shows also that some quite eminent people, even within the Catholic Church, were willing to openly support his attempts to develop a new and more accurate model of the cosmos. In fact, what is maybe even more surprising is that the private secretary to Pope Clement VII, Cognominatus

79. Pope Clement VII heard a talk given by his private secretary on Copernicus's theories during 1533, in the Vatican gardens. He rewarded his secretary with a book testifying to his pleasure in the event.

Lucretius, actually gave a talk about Copernicus's theories three years earlier in the Vatican gardens. There is written evidence of this on the title page of the book Cognominatus Lucretius was given by the pontiff in thanks. Copernicus may never have known of this event, but even so, Cardinal Schoenburg's letter should have assured him of the support he had even in the highest quarters; nevertheless Copernicus himself was still worried.

Conscious of the controversy his theory was likely to cause Copernicus spent a great deal of time checking and rechecking the observations he had made in an attempting to avoid any kind of criticism. It also meant rechecking his interpretation of other astronomers' material too – as many of his calculations were based on the work of long-dead scientists and philosophers like Ptolemy, whose observations Copernicus was using in support of his own theories. In addition it involved ensuring that the mathematics he used to back-up his ideas were correct too and could not be challenged. Such concerns may have been one of the main reasons why Copernicus had delayed publication of his work.

There were, though, other problems preventing him resolving some of the difficulties he was having tying in his astronomical observations with his proposed new model for the universe. Copernicus was trying to show that the earth was just another planet and not the centre of the cosmos, but he was still being hampered by his belief that what he thought of as God's creation should be anything but perfect. He imagined the orbits of the planets as being perfectly circular but his observations of their movements suggested otherwise. He could see for example that their distance from the sun did not appear to remain constant. At different times their orbits took them further away from the sun and at others nearer.

The truth is that the orbits of the planets are not circular. We know today that they are elliptical: more akin to an oval than a circle. Copernicus could not accept this. In fact he fell into the same trap as Ptolemy. Ptolemy had introduced epicycles into his earth-centred model of the universe to account for their retrograde motions and explain why the planets so often appeared in a different position to what would be expected in a geocentric system. Epicycles were thought of by Ptolemy as small circular movements that the planets made that caused them to swing towards and then away from the earth as they orbited around it. Copernicus was faced with a very different problem but nevertheless opted for the same solution. He introduced the concept of epicycles to explain why the planets' movements did not conform to what one would expect of bodies following perfectly circular orbits around the sun. Because of this one false assumption Copernicus's theory became entangled in some very complicated mathematics simply to explain the inconsistencies between his hypothesis and what could be actually seen when observing the planets. There is little doubt that this too contributed to the delay in publishing his ideas.

Kohler, in his book about the astronomer, refers to Copernicus as the timid canon – and with some justification. Copernicus may have been very forward and quite courageous in his thinking, given the times in which he lived; but he was certainly reticent about expressing his opinions on the public stage. His hesitation to publish was partly an expression of his fear of criticism rather than actual persecution; but was probably also due to the fact that his calculations had become mired in uncertainties that dogged him to the extent that he had even began to doubt his own theories.

A further problem was that, unfortunately Copernicus had relied far too heavily upon the earlier observations of other astronomers, including those of Ptolemy whose own model of the universe he was challenging and other even more ancient sources such the observations of Chaldean astronomers. He had then realised rather belatedly that errors had crept into his own calculations as a result. Consequently, from 1537 onwards Copernicus made more observations than at any other time in his life. He was facing the daunting task of reworking much of what he had done, or maybe even having to start again from scratch. Under the circumstances the idea of publishing his theories seemed ever more remote, but help was at hand.

In 1539 Copernicus had a visitor. George Rheticus was only twenty-five years old but had already been appointed as Professor of Astronomy and Mathematics at Wittenberg University whereas Copernicus at the time was sixty-six years old. Rheticus had read one of Copernicus's papers and had been impressed. Consequently, he was curious to meet the ageing astronomer. He had intended to stay only a few weeks when he first arrived but, in fact, Rheticus became so interested in Copernicus's work he actually ended up staying for two years.

In one way this was really quite surprising as George Rheticus was a follower of Martin Luther the reformist priest who had broken away from the established Catholic Church of Rome, in other words George Rheticus was what we now call a Protestant although that term had not yet been coined then. He was actually risking his life visiting a Catholic country like Poland. In fact Rheticus' visit was extremely daring as it coincided with an edict issued by Bishop Danticus that banned all Lutherans from entering Ermland, threatening confiscation of their goods or even execution. Copernicus too could have brought down upon himself the wrath of the Church in Rome, as well as that of his own bishop, for consorting with what they would have seen as a follower of a heretic priest who was an enemy of the Catholic Church. Rheticus though was a born rebel. Indeed trouble seemed to follow Rheticus as he himself was to later lose his eminent position as a professor of mathematics following a rather unpleasant, drunken homosexual incident involving a student at the university where he was employed. Regardless of the outward propriety of the times, both bisexual and homosexual behaviour were not only commonplace during this period but possibly also more widely tolerated in some ways than today but apparently George Rheticus, typically, did not know where to draw the line.

Copernicus, with the assistance of Rheticus, set about checking and refining his calculations. We know nothing about the personal relationship between the two men apart from the fact that they quickly became good friends, with the usually reticent and cautious astronomer giving Rheticus full access to all his manuscripts. Rheticus on his part gave his all to the task in hand – and a daunting task it was.

Unfortunately, falling into exactly the same trap as both Ptolemy and Copernicus had, Rheticus found himself also struggling with the problem of epicycles that had defeated the older astronomer. Despite his best efforts he was unable to help Copernicus formulate a planetary system that accurately corresponded with their astronomical observations and was instead forced to frustratedly introduce an increasingly complex series of epicycles in an attempt to make the movements of the planets fit with the concept of circular orbits. These we now know are in reality ellipses not circles, creating problems that were insurmountable even for a brilliant mathematician like Rheticus. The strain nearly proved too much for the young man who by one account may well have had a minor mental breakdown at one point. There are descriptions given later by astronomers like Johannes Kepler of Rheticus banging his head against the wall in frustration.

Eventually though, by introducing more epicycles, Copernicus and Rheticus were able to produce mathematical tables that were sufficiently accurate to support Copernicus's hypothesis of a sun-centred planetary system. But despite the encouragement of both his old friend Bishop Tiedemann Giese and his disciple, Rheticus, and even the letter from Cardinal Schoenburg that implied the tacit support of the pope himself, the stubborn old man would not publish.

The turning point probably came during a visit Rheticus and Copernicus made to Loebau Castle, the official residence of Bishop Tiedemann Giese where they stayed with Copernicus's long-time friend. For a short time it must have been one of the most enjoyable periods in Copernicus's life being in the company of two friends who understood and supported his work. Nevertheless there was a serious task to be accomplished and both the other men spent much of their time together trying to persuade the ageing astronomer that he should finally consider publishing his ideas. Copernicus, though, was very conscious of the fact that he was about to turn all the ideas people had previously held about the universe on their head and he knew that he had to be certain that his data stood up to scrutiny. Indeed it is said that one of his critics, the reformist Martin Luther, actually referred to Copernicus as, 'That fool who would turn the world on its head'.

We know from what George Rheticus wrote later, that to avoid causing controversy, Copernicus had considered publishing only the tables that were derived from his new astronomical theory, without any explanation of the underlying principles upon which they were based, concerning the motions of the earth and other planets. He reasoned that ordinary people would then remain unaware that he was proposing a sun-centred cosmos to replace the accepted geocentric system, whereas the true mathematicians and others with sufficient insight, would still be able to divine the real nature of the model represented by the tables for themselves. This was very much in keeping with similar esoteric practice of sects like the Pythagorans and others who had in the past sought to limit the knowledge they gained to a selected few.

Both Rheticus and Tiedemann Giese were however against this subterfuge. It was his long-time friend the prelate who eventually persuaded him to change his mind, telling him that it would be an incomplete gift to the world unless he explained the reasoning behind the tables as well as the foundation and proofs upon which they relied. Naturally he was well aware that pursuing this policy would ensure that there could be no misunderstanding regarding the true nature of the planetary system Copernicus was suggesting. Despite Copernicus's own misgivings, Bishop Giese and Rheticus, exasperated by the ageing astronomer's continuing reticence, both wanted the cat let out of the bag and into the public domain.

The outcome of these deliberations between the three friends was a compromise. Although he himself did not want to risk the public ridicule that might have followed if his theories were rejected, or were found flawed, Copernicus did allow George Rheticus to write an initial, popularised account entitled *Narratio Prima* (First Account) outlining his ideas for a heliocentric system: helios being the Greek name for the sun. In an attempt to deflect criticism it was decided that it should be written in the form of a letter from Rheticus to his old teacher, Johannes Schoener, describing the ideas of his new mentor and it was mutually agreed that Copernicus's name would only be mentioned once on the title page and that elsewhere he would be referred to only as *domine praeceptor*, or learned teacher. The full title of the manuscript Rheticus wrote was *Narratio Prima de Libris Revolutionum*. The English translation would be, *The First Account of the Books of Revolutions*, and it described only the last of the four as yet unpublished books explaining Copernicus's ideas.

Within ten weeks of his arrival in Frauenburg, in the summer of 1539, George Rheticus had completed the *Narratio Prima*, the first account of Copernicus's work, presented it to the printers in Danzig and in only a few months, by February of the following year, had seen it through to publication. Copies were then distributed to a number of interested parties. Among these was Duke Albert of Prussia the erstwhile, grand-master of the Order of Teutonic Knights, who had once been the sworn enemy of Copernicus's maternal uncle Bishop Lucas Waczenrode. Losing the war against the Polish king had broken the power of the knights and the grand-master, renouncing any rights to Ermland and accepting in return his new title as duke, had then become a Protestant: effectively striking the death-knell for the order.

Further copies of the *Narratio Prima* were dispatched to various other influential luminaries in both Protestant and Catholic regions of Europe. Cardinal Schoenburg, who had written the much treasured letter to Copernicus, received one for example. There was even a second edition that was printed by one admirer in Basle, spreading the word even further. Many of those who had read the *Narratio Prima* now joined his friends in urging

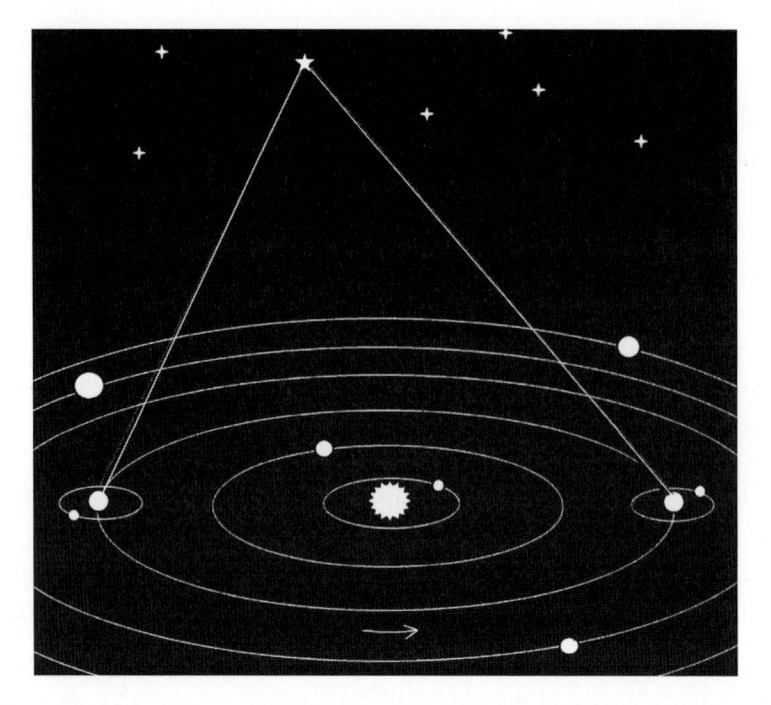

80. The solar-centred planetary system Copernicus was proposing implied that the earth made a vast orbit around the sun each year. It was therefore assumed that if this was the case the stars should naturally appear at different angles at various times of the year as the earth moved through space. Attempts were made to measure the parallax of particular stars from opposite sides of the earth's proposed orbit; but no such difference could be detected! There were only two possible explanations for this: either Copernicus was wrong or the stars were so far away that using the instruments of the day the two lines appeared to all intents and purposes parallel. The angles were simply too small to measure but instead of accepting the true explanation many people simply thought that this proved Copernicus's theory to be wrong.

Copernicus to publish the complete work. It would seem that he now had no legitimate reason to fear ridicule or censure. Nevertheless, the ageing astronomer was himself still hesitant about publishing his theories in full.

There was though one other particularly serious and persistent problem that may have caused Copernicus to still hesitate before presenting his theory to the world. This was to do with the method of calculating the distance of remote objects such as celestial bodies like the moon or the planets that involved measuring the parallax. This principle was mentioned in an earlier chapter, describing how Copernicus and his tutor, Professor Domenico Maria Da Novara, had calculated the moon's parallax while he was studying at Bologna University. The reader may remember that they had tried to calculate the distance of the moon from the earth by comparing the difference in the angle, at different phases.

The same principle had been applied in a different form to measuring the distance of the stars too but without result. It was found that there was no discernable difference in the position or angle at which a star appeared from one location to another. This it was discovered was even the case when taking readings from opposite sides of the earth's orbit. It was realised, quite rightly, that this could be accounted for if the stars were very distant objects, which would simply make the angles at which they were seen from different parts of the earth too similar to show any measurable variation. These very observations, though, might also be interpreted in another way, to suggest that the earth was stationary as Ptolemy had claimed and not orbiting the sun as Copernicus believed it to do. It was reasoned that if the earth made a vast orbit around the sun then the stars should naturally appear at different angles at various times of the year as the earth moved through space. But no such difference could be detected! There were only two possible explanations for this: either Copernicus was wrong or the stars were far more immensely distant objects than anyone had imagined, which could mean that any discrepancy in the angles was too small to measure even at opposite sides of the earth's orbit.

We now know that the second explanation was in fact the true one; but people during the Middle Ages just did not have any experience or concept of such vast distances. This must have worried Copernicus and indeed it remained one of the main stumbling blocks to the wholesale acceptance of the Copernican model of the universe by some scientists for more than a century after the astronomer's death. Although more than 300 years were to pass before it actually became possible to measure the distance of the stars accurately enough to detect any parallax, it nevertheless gradually became obvious to later astronomers that Copernicus's concept of the cosmos was correct. This in turn made people realise that the universe had to be far more enormous than ever before envisaged. In this respect Copernicus really did change people's concept of the universe in which we live and along with his concept of a sun-centred planetary system, it must remain one of his major achievements.

81. The city of Nuremberg was once the capital of the Holy Roman Empire. Representing a group of trading nations, the empire was the medieval equivalent of the European Common Market. It was also one of the cultural capitals of medieval Europe and it was to printers in Nuremberg that Copernicus's book, containing his revolutionary theory about the movements of the planets, was finally sent for publication.

XVI

THE REVOLUTION OF THE SPHERES

It was really Rheticus, along with Tiedemann Giese, who finally persuaded Copernicus to publish his theory that it was the sun and not the earth that the planets revolved around. It may have been George Rheticus's preliminary explanations in the *Narratio Prima* that finally prompted Copernicus to act and by 1542 his great lifetime's work was completed and ready for publication. It was to be entitled *De Revolutionibus Orbium Coelestium Libri VI* or in English *Six Books Concerning the Revolution of the Spheres.*

Over the years Copernicus in one way or another had lost all of his friends and he must have become very lonely and isolated in the latter part of his life. His brother Andreas had become sick and died decades before and his friend and fellow canon, Tiedemann Giese, was now Bishop of Kulm and lived in residence many miles away. Canon Sculteti had been expelled from the chapter and his housekeeper and cousin Anne Schilling was forced to leave Frauenburg. When the time finally came to publish his book his friend Rheticus too had to leave. He had been offered a new post in Leipzig and had to take up his duties as Professor of Mathematics at the university there. Copernicus was now to all intents and purposes alone. In 1542, as the year drew to a close he also suffered the first of a series of strokes. He was by now bedridden and, when in most need, without even the company of his erstwhile friends.

At the beginning of the new year Bishop Tiedemann Giese was sufficiently concerned to write to one of the other canons, George Donner, who was also

Orbit of inferior planet

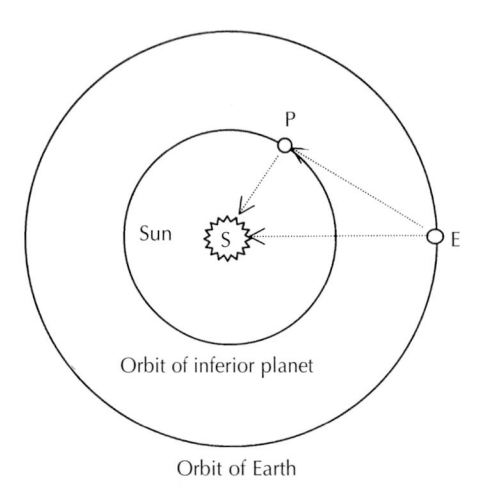

Orbit of Earth

82. When inner (inferior) planet P moves directly away from the earth at what is known as maximum elongation, it will appear practically stationary to the observer. One can deduce from this configuration that at this point when seen from the inner planet, the earth and the sun would be at right angles to one another. A second angle between the two observations taken from the earth, of the sun and an inner planet can be easily measured directly by the observer. With these two pieces of information one can calculate the ratio of the inner planet's orbit in relationship to the earth's orbit by comparing the radius SE to the radius SP in a right angle triangle.

To compare the orbit of an outer (superior) planet with that of the earth, one has to first record the precise time that the sun, earth and the outer planet were directly in line. Another observation is then made when the earth is moving directly away from the outer planet: at this point the other planet appears practically stationary and one can check directly, that the angle between the observed position of the sun and that of the outer planet is a right angle. A second angle has to be determined to calculate the outer planet's orbit. This can be done by one of the most reliable pieces of information available to an observer, which is the time taken by the earth to orbit the sun. If one assumes that orbit is circular and the earth's speed constant (neither of these is actually true but the method is still basically sound despite the small variations), by measuring the time that passes between the two observations one can tell what segment of its orbit the earth has travelled in the intervening period. From this one can deduce the degree to which the angle would have changed if it were possible to view the earth from the sun. With these two angles one can find the difference in ratio between the radius SP of the outer planet's orbit and that of the earth SE.

Despite being able to compare the size of any planet's orbit with that of the earth, Copernicus could still not determine the actual size of any orbit without first calculating how far the sun was from the earth: which he needed to know in order to find the diameter of the earth's orbit. Knowing the approximate size of the earth, medieval astronomers did try to estimate its distance from the sun by measuring the parallax of the sun. This method has been described earlier in the book as a way of assessing the moon's distance; but as the sun is much further away the results obtained were grossly inaccurate.

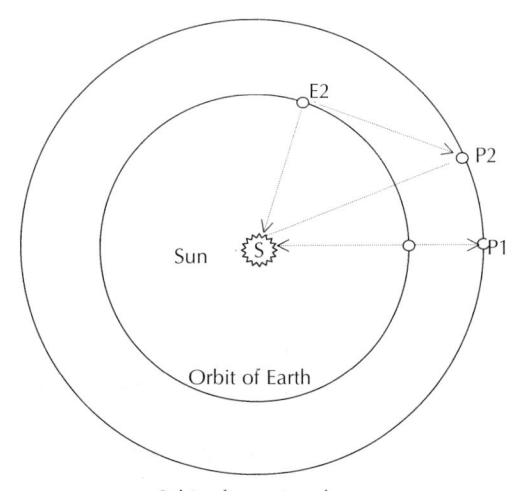

Orbit of Earth

Orbit of superior planet

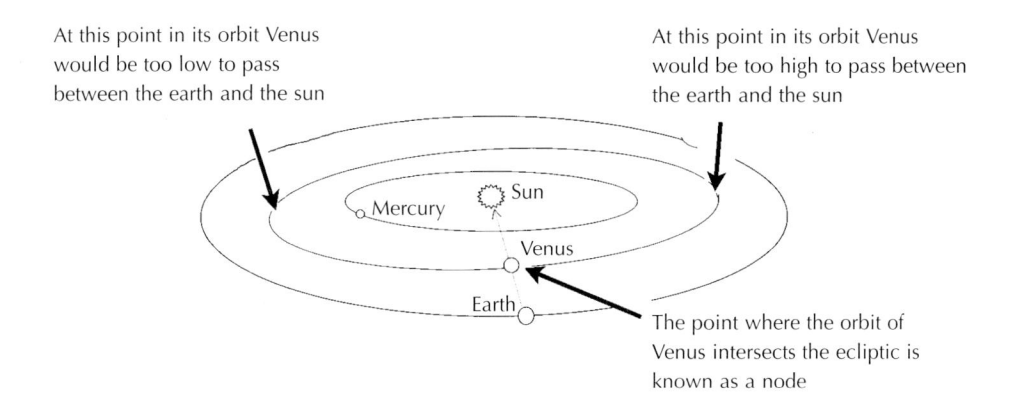

At this point in its orbit Venus would be too low to pass between the earth and the sun

At this point in its orbit Venus would be too high to pass between the earth and the sun

The point where the orbit of Venus intersects the ecliptic is known as a node

83. Venus can only be seen passing across the face of the sun when several conditions are met. The first is that the earth, Venus and the sun are all in line, to produce what is known as an inferior conjunction. The second requirement is that this happens at a point where the orbits of two planets coincide. This is not always the case because the orbits of the two planets are not exactly in the same plane. A third condition is that this occurs where the orbit of Venus crosses on the plane of the ecliptic: marking the apparent path of the sun across the sky. Moreover, a transit of Venus across the face of the sun can obviously only be seen during the day, therefore those on the night side of the earth at the time it occurs, will not be able to observe this phenomenon. This specific combination of circumstances means that transits of Venus occur in pairs about eight years apart; but this is usually followed by a gap of more than a century before another transit can be observed. Only one of the two transits of Venus that occurred in Copernicus's lifetime was during daylight hours in Europe but until the invention of the telescope it would have been impossible to see against the glare of the sun. Even then the image from the telescope would have to be focused on a screen to be viewed, for the observer to avoid being blinded by the sun's rays.

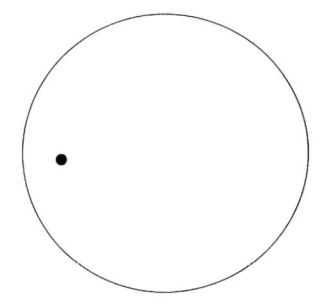

Whereas for the first observer the planet Venus has already completed part of its journey across the disk of the sun, for the second observer it has barely started even though they are both observing the transit at the same time.

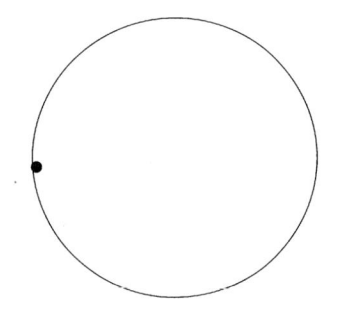

To observers in different locations the transit of Venus would appear to begin and end at a slightly different time. The amount that this differed would allow one to compute how far Venus was from the earth; and as the ratio of the earth's orbit to that of Venus is known, the sun's distance from the earth can also be determined. But the first astronomer to realise that this phenomenon could be used to assess the sun's distance was Edmund Halley in 1679, long after Copernicus's lifetime.

84. In 1543, the year that Copernicus died and his book *The Revolution of the Spheres* was finally published, another important work made its first appearance too. This book was written by a doctor, Andreas Vesalius who was the first modern anatomist. It was a sensation, unlike Copernicus's lifetime's work that was largely ignored by the population as a whole. Vesalius's book was illustrated with examples of dissections and autopsies carried out by Vesalius. Thus they were remarkably accurate unlike previous anatomical illustrations. Ultimately, although the influence of Copernicus's theories was not immediately apparent both books were to have a huge impact and each in its own way helped in the renaissance of their respective disciplines and formed the basis for modern scientific thinking.

resident at Frauenburg Cathedral, asking him to keep an eye on the ailing Copernicus. During the first few months of 1543 Copernicus's condition became worse. He suffered further strokes and a heart attack – it was obvious to everyone that the old astronomer was dying. In the meantime the book that represented his entire life's work was being prepared for publication.

But even here all was not well. When Rheticus took the manuscript of Copernicus's book, he went first to Nuremberg to arrange for it to be printed before journeying on to Leipzig. Unfortunately, the responsibilities of his new post meant that as he was unable to stay and see the project through to completion, instead he handed over the responsibility for supervising the publication of the book to another friend of Copernicus, Andreas Osiander, who was an important political figure in the city. But Osiander was also a theologian and had real doubts about the reception such a controversial book might receive from the Catholic Church.

Osiander was worried that the Catholic Church might condemn out of hand, any publication putting forward a theory suggesting that the earth was not the centre of the cosmos. These concerns led him to betraying the trust of both Rheticus and the ageing Copernicus by replacing the original preface that Copernicus had written with one he had written himself. The preface Copernicus had written was in the form of an address to Pope Leo III, in which he had explained both his reasoning and his motives for writing the book to the pontiff.

Andreas Osiander's substituted preface was very different and falsely claimed that the book did not seek to prove that the earth and other planets really orbited the sun at all but merely used this notion as a means of constructing a more reliable mathematical model for predicting astronomical events rather than representing the universe as it really was. Whereas Copernicus had been relying upon the mathematical calculations he included in the book to prove the logic of his arguments, Osiander claimed that they were merely a convenient contrivance. In other words denying the very thing that Copernicus was hoping his book would prove after dedicating an entire lifetime's work as an astronomer and mathematician.

When he discovered the fraud Rheticus was furious, but Copernicus was probably spared any distress this betrayal might have caused him. Having suffered a heart attack and a series of strokes by the time the book was printed he was dying. When the book he had worked so hard to complete was put into his hands for the first and only time he already lay on his deathbed. His mind by then was already befuddled and his sight was failing and it is doubtful that he would have seen that any changes had been made. One may assume, therefore, that he probably died still believing that his idea of a heliocentric cosmos was finally going to be presented to the world as he had always meant – as a genuine theory that would change people's concept of how things truly were and the place of planet earth in the universe.

Rheticus wrote to Copernicus's other great friend Bishop Tiedemann Giese shortly after Copernicus had died, describing how the old astronomer, who had been in a coma, briefly regained consciousness when the book was put into his hands. We know from his comments that Rheticus was convinced that Copernicus was sufficiently aware to realise that it was his book that he held but was probably unable to see what was written in it – and thus blissfully unaware of the substitution that had been made by Andreas Osiander.

Strangely enough despite everything, the final irony was that Osiander's substituted preface did nothing to prevent other astronomers and mathematicians recognising the logic and strength of Copernicus's reasoning and perceiving what really his true intentions were when he wrote *The*

85. When the book he had worked so hard to complete was put into Copernicus's hands for the first and only time, he already lay on his deathbed. His mind by then was already befuddled and his sight was failing and it is doubtful that he would have seen that any changes had been made. He probably died still believing that his idea of a heliocentric cosmos was to be finally presented to the world (adapted from *Vie des Savants Illustres,* Figuier, Louis, Hachette 1883).

Revolution of the Spheres. Moreover, at the same time, Copernicus's critics in the Church were disarmed by Osiander's presentation of the heliocentric model as no more than a convenient mathematical contrivance. Indeed it could be argued that it may in reality have merely been Osiander's intention to protect Copernicus from censure and persecution all along.

Many others who read his book may in fact have originally believed that Copernicus himself had written the preface; but only to disarm his critics in the Church and elsewhere. Eventually the truth became known anyway and Copernicus's heliocentric theory was widely accepted by the academics throughout Europe. In this respect Osiander's subterfuge may have had a positive outcome by preventing the Church banning the book until well after the ideas it contained were known and fully publicised. Ultimately, the truth was out and the Catholic Church could not put the genie back in the bottle; but that is not to say they did not try.

Consequently later astronomers, regardless of what they personally believed, were at least familiar with the heliocentric model that Copernicus had proposed and most accepted it in preference to an earth-centred cosmos or any other of the earlier theories. There was though still resistance by many in the Church to any idea that threatened to challenge the pre-eminence of the earth as central within a universe that was perceived as created by God. Following the death of Copernicus the Church became increasingly less tolerant of critics and those that questioned established beliefs or Church teaching.

At the time when Copernicus was born European culture was beginning to undergo a number of profound changes. The Middle Ages had been to some extent a time of stagnation. The occupation of Spain by the Muslims led to the introduction of Arabic numerals and more advanced forms of mathematics and geometry into Europe: providing the tools by which scientists could achieve a better understanding of our universe and the world in which we live. The fall of Constantinople also contributed to the reintroduction of earlier wisdom gained from manuscripts that had originally come from the cultures of ancient Greece, Egypt and the old Roman Empire, as did the invasion of Muslim-controlled Sicily by Norman adventurers.

Just as importantly, Copernicus lived in a part of Europe where there was far more tolerance than elsewhere and Humanist philosophies that had arisen as the result of this influx of new ideas were beginning to encourage broader, less critical attitudes that permitted questions to be asked that previously would have led to censure and persecution.

In fact, a number of peculiarly fortuitous circumstances probably contributed to Copernicus's eventual success. To begin with he had been fortunate enough to live through the reign of several benevolent popes and even when they may not have agreed with his theories they had recognised and valued his skills and up to a point were willing to respect his right to express his own opinions even though he was a canon of the Catholic Church. Moreover, his abilities as a mathematician and astronomer were actively sought by the Church to help with the much needed reform of the Julian Calendar. The fact that he never got around to doing it in retrospect seems quite unimportant given his actual accomplishments.

Copernicus also had friends from both the Catholic Church and the new Protestant faith who helped and encouraged him. He had a powerful patron and protector in his uncle and an independent income for the rest of his life as result of the position Bishop Lucas Waczenrode obtained for him as a canon at Frauenburg Cathedral. And who knows if his father had not died when he did, Copernicus would not have been adopted by his powerful uncle and may have simply have followed in his father's footsteps and become a copper merchant instead; not the astronomer who moved the earth and stopped the sun.

Finally it was only due to the young professor of mathematics, George Rheticus, that Copernicus was persuaded to publish his theories when he did. In fact the strokes that were to cause his death occurred very shortly after he completed *The Revolution of the Spheres*, which is why Copernicus only actually saw a copy of his book while on his death-bed. For this reason if no other, we and Copernicus owe a great deal to George Rheticus for ensuring that the heliocentric system was eventually accepted; but strangely his efforts were never acknowledged by the astronomer himself in *The Revolutions*. Nowhere is there even the least mention of Rheticus; nor any reference to the essential part he played in helping Copernicus correct and rework his calculations, organise the manuscript, or in getting the book published.

Moreover, the effort expended in checking and correcting Copernicus's calculation had nearly cost Rheticus his sanity. He must have been terribly hurt by this unforgivable omission. It is difficult, if not impossible, to understand why this happened and Copernicus without a doubt denied his young collaborator the public recognition he so richly deserved. Tiedeman Giese wrote to him trying to reassure him that the omission was due to Copernicus's age and confused thinking, but this kindly attempt to ease the younger man's sense of betrayal was hardly supported by the facts, as the astronomer despite his age had been perfectly lucid at the time of preparing the manuscript as Rheticus must have been fully aware.

This may explain why Rheticus, despite his original enthusiasm for Copernicus's heliocentric model of the universe, took little part in promoting the astronomer's work after he had died. He seemed to lose all interest in the subject of astronomy too. Soon afterwards his career also took an abrupt downward turn when he lost his position as professor of mathematics following a drunken homosexual assault upon a young student to whom he was tutor. He never recovered from this abrupt and disastrous fall from grace and spent the rest of his life eking out a wretched existence as ordinary physician rather than as a mathematician or an astronomer.

86. In 1600 after subjecting him to a long imprisonment and eight years of torture the Catholic Church burnt one follower of Copernicus's theory at the stake. Canon Giordano Bruno went even further than Copernicus: he suggested that the stars were distant suns that may have their own planets orbiting around them. He also suggested that people might exist on these planets with their own version of Christ and the Christian religion. This was too much for the Catholic Church who would not countenance any challenge to their pre-eminence as the 'one true church' and although Bruno managed to escape their immediate wrath by fleeing into hiding, he was eventually tricked into returning to his native Italy and fell into the clutches of the Church.

XVII

THE COPERNICAN REVOLUTION

His great work on the movement of the planets was only published after he had died and eventually it was not what Copernicus achieved during his lifetime that was to make the real difference to our thinking but what occurred after he was gone.

Ultimately, there is no doubt whatsoever that the ideas contained in Copernicus's book, *The Revolution of the Spheres*, were to totally change the science of astronomy; but what is sometimes not realised is that Copernicus's new planetary system was neither any simpler in its detail than that of Ptolemy, nor were the tables it produced plotting the movements of the planets any more accurate. Because he, along with every other astronomer at that time, believed that the orbits of the planets had to be perfectly circular, Copernicus had to employ nearly as many epicycles as Ptolemy before the observed movements of the planets could be predicted using his planetary tables and if anything the results from Ptolemy's tables were more reliable; even if based upon a false premise. Neither was the idea of a sun-centred system new: Aristarchus had suggested the same thing nearly 2,000 years before.

In fact after a lifetime's effort Nicholas Copernicus was saying no more than that the movements of the known planets could be more easily explained if we assumed that the earth went around the sun and not the other way around. He knew that he had not proved this was actually so: in fact he was very conscious of the extent to which he had failed in this endeavour. It was one of the main reasons why he was so hesitant about publishing his ideas. He was not even sure that the theory was true and he was well aware that his calculations were flawed. This was partly due to his slavish reliance on the work of earlier observers without checking their work for errors. It is known that he made very few observations of his own. A further problem was that Copernicus could not reconcile the belief he erroneously held that the orbits of the planets were perfect circles, with what he observed in their movements, without resorting to the reintroduction of the very epicycles he had sought to eliminate from the planetary tables.

This was one of the major stumbling blocks that delayed the completion of his thesis and it was just one of the things that still remained disconcertingly unresolved when he died. This false assumption that the planets' movements could only be expressed as circles had dogged the efforts of every astronomer before Copernicus and many of those who followed him.

This led to errors in Copernicus's calculations that could only be corrected much later, once the astronomer Kepler finally realised that the data he himself had accumulated on the orbit of each of the planets actually represented ellipses, and that this described the planets' true path around the sun and not a circle as everyone had previously assumed. Even then physical proof of Copernicus's theories of planetary movements was only provided decades after his death, when Galileo first began observing the planets through the newly invented telescope.

Nicholas Copernicus had his adherents even before *The Revolutions* was published, and despite its flaws and obvious shortcomings, the Copernican system was still to gain acceptance among some of the most eminent and able astronomers well in advance of any proof of its veracity. Ultimately what made it acceptable was its usefulness as a working model, which came from the ease by which its basic hypothesis could be applied to interpret particular phenomena. Simply by assuming a sun-centred system for example, the retrograde movements of the planets immediately became comprehensible. The system was in such respects more logical than any other previously proposed and, moreover, it possessed features that would appeal to the mathematician as much as the cosmologist – which were its underlying simplicity and elegance.

Copernicus himself identified one of its other main strengths in his preface, which was that unlike the Ptolemy's geocentric system, all the elements were interdependent and an integral part of the whole: it would only work if each of those elements remained unchanged. This means that the values and relationships between each part of the system are predictable, which alone makes certain things possible that before would not otherwise have been practical to attempt. Just one example is calculating the distance between the earth and the sun.

The distance from the earth to the sun is important because it is used as the basic unit of measurement when calculating other distances in astronomy. For this reason it is known as an astronomical unit, or AU for short. But in the time of Copernicus no-one had yet managed to calculate this distance accurately and Copernicus himself was equally unsuccessful. In fact the correct distance was not discovered until after the telescope was invented. Nevertheless, it is only by recognising that the earth is part of a heliocentric planetary system that the true value can be found. There are at least two ways for example that Copernicus himself could have attempted to estimate the distance between the earth and the sun; but apparently he never realised it was possible. It is also doubtful if he ever had the opportunity to do so under the very specific conditions that would have been required.

One method would have been by calculating the parallax during a transit of Venus, when the planet passed directly in front of the sun. This would involve making two observations, measuring the movement of the planet across the face of the sun from two different, widely separated locations. The transit would appear to begin and end at a slightly different time at the two observation points. The amount that this differed would allow one to compute how far Venus was from the earth; and as the ratio of the earth's orbit to that of Venus is known, the sun's distance from the earth can also be determined. It is not even necessary to make the observation simultaneously as providing this is allowed for in the calculations the distance can still be computed from the angular difference. The first astronomer to realise that this phenomenon could be used to assess the sun's distance was Edmund Halley in 1679, long after Copernicus's lifetime. Transits of Venus practically always occur in pairs eight years apart; but unfortunately rarely, with each pair of transits being separated by more than century. Nevertheless, during his lifetime there were two transits but the first, which was in June of 1518 occurred during the night. There would have been an opportunity to observe a transit of Venus in the June of 1526, while Copernicus was living and working as a canon at Frauenburg Cathedral but he would have needed a telescope and this device was not destined to appear upon the scene until the next century.

Using the transit of Venus to assess the sun's distance is essentially the same method that both the ancient Greek astronomer Aristarchus attempted to use in the third century BC and Hipparchus employed in the second century BC to discover how far the moon was from the earth. Their calculations had relied

86. It was the later astronomer Kepler who eventually resolved one of the problems that had most frustrated Copernicus and nearly driven his friend George Rheticus mad with despair. Whereas Copernicus and Rheticus had both assumed the paths of the planets around the sun must be circular, Kepler was able to show that their orbits were in fact elliptical. This not only explained the still unresolved inconsistencies that existed in Copernicus's model, but also allowed astronomers to finally dispense with the now completely unnecessary complication of introducing epicycles into their calculations, that had been used previously to explain the apparently erratic movement of the planets.

upon comparing the amount of the sun visible during a solar eclipse, at two different locations about 1,000 kilometres apart. Aristarchus also attempted to discover how far away the sun was by similar means, but the vast distances involved combined with his own flawed observations meant he was unsuccessful.

A second method that the ancient Greek astronomers, or Copernicus could have used, would have been by calculating the distance of another nearby planet such as Mars at the point at which it passed closest to the earth; although preferably by recording the changing parallax as the earth rotated rather than taking observations from two different locations. This would also give the earth's distance from the sun as the ratio of earth's orbit to that of Mars is also known. In fact it was more than 100 years after Copernicus died before the sun's distance was measured with any accuracy, using similar methods to those described.

Nevertheless, it was Copernicus's concept of a sun-centred planetary system that actually made such calculations possible. In fact nearly everything that we have discovered about the universe since the time of Copernicus is the result of the new cosmology he engendered. Ptolemy's and Aristotle's models of the universe were not only flawed, beyond a certain point, they were unworkable. Even though they produced data to help predict celestial events with remarkable accuracy, to state the obvious, they would have severely limited any real understanding of the universe in which we live.

It is strange considering the immense influence that Copernicus had upon our concepts and understanding of the universe, that in some ways he appears to have actually been rather dilettante about his research: there were for example long periods in which it appears he did nothing – proving that the earth and other planets circled the sun was in effect his pet project – but what he did achieve once he was persuaded to publish, proved to be ultimately of fundamental importance to the development of astronomical science.

Although the theories of Copernicus were to turn medieval ideas about the nature of the universe on their head and change forever the sciences of astronomy and cosmology, he was by nature a very conservative man. Kohler in *The Sleepwalkers* refers to him as the timid canon. In his life Copernicus had but that one single hobbyhorse to ride, which as we have seen he did cautiously – and fearing ridicule – often surreptitiously. The ideas he had discussed among trusted friends and broadcast to only a very limited number of confidants and fellow academics, were nevertheless well known to many others, including those like the reformist Martin Luther who both disparaged and disapproved of them. He also had his supporters though even in the Roman Catholic Church, like Cardinal Schoenburg as well as his adherents among the Protestant reformists. In other words it is obvious that there were plenty of people who were aware of his basic hypothesis: suggesting a sun-centred planetary system.

This was so, long before the book was published, to explain in detail the mathematics and concepts behind his thinking. But publication of *The Revolutions* was still an important event even though it occurred on the day of his death, because it made a public statement about his commitment to the heliocentric model of the universe which he had so long championed. It was not even necessary that people read it – and very few did – what was significant was that despite this, everyone who mattered knew what Copernicus had said. Because of Osiander's misleading preface and Copernicus's own somewhat apologetic style of presenting his thinking, it did not cause the stir one might imagine among the conservative elements of the Church and therefore *The Revolutions* was not officially banned by the Church for nearly sixty years: by which time the genie was well and truly out of the bottle.

There may have been several reasons that caused Copernicus to hesitate before publishing *The Revolution of the Sphere*: such as a fear of ridicule, or the need he felt to constantly to check and improve on his calculations to avoid criticism. But he may have had other more tangible fears of persecution too. Osiander's substitution of the original preface was to some extent proof of this. Such cautiousness may well seem unfounded given the support he had already received, from people like Cardinal Schoenburg and his friend Bishop Tiedemann Giese; but despite his powerful connections, one should remember that he had never before publicly challenged the old earth-centred concept of the cosmos promoted by the Catholic Church, which he as a canon was meant to serve.

In this respect we can certainly be grateful that Copernicus had lived and worked before the iniquity of the Inquisition had made itself fully felt in Northern Europe. The Inquisition originated in Catholic Spain. Originally a Christian country, Spain had been conquered and ruled by Muslims for several centuries until they were finally overthrown and most, apart from a few enclaves of resistance, were expelled from the country a short time before Copernicus had been born. A century after their defeat and the final demise of Muslim rule in Spain, the Inquisition had been formed. Muslim rule if nothing else had been tolerant of other religions; but this was far from the case once Christian sovereignty had been re-established and the Inquisition was the embodiment of the worst religious prejudices and may have even been a belated response to the previous Muslim domination of Spain.

The word '*inquisition*' means investigation. The Inquisition basically consisted of a group of priest and lay people who sought out anyone who they thought opposed the teaching of the Church or questioned Christian beliefs.

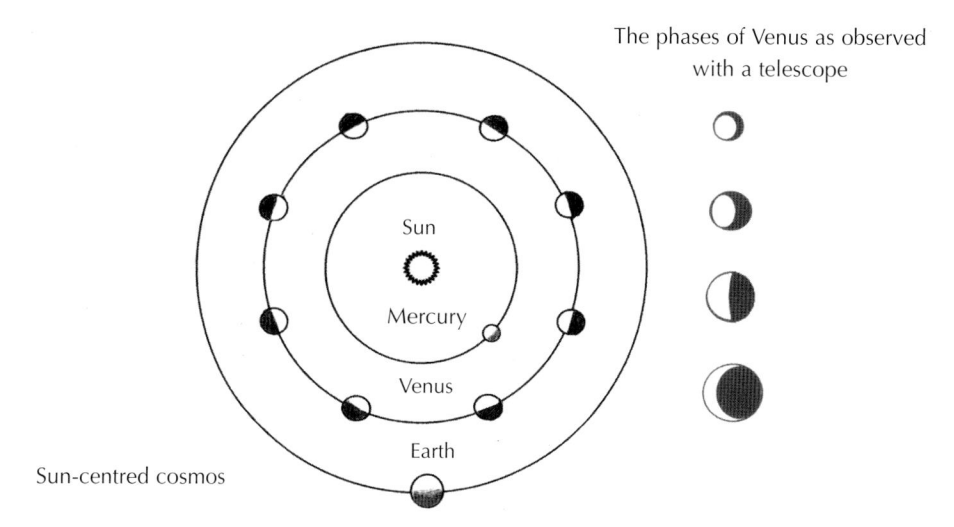

The phases of Venus as observed with a telescope

Sun-centred cosmos

87. Ptolemy believed in a geocentric cosmos; but to explain the fact that both Venus and Mercury were only seen in proximity to the sun Ptolemy therefore had to assume that these two inferior planets closely followed the sun as they and the sun all orbited around a stationary earth. Moreover, to account for the observed easterly and westerly movements of the two planets he also had to introduce the idea of epicycles, in which planets followed additional smaller circular paths as they orbited the earth. It was not easy to refute this theory by naked eye observations alone.

It was more than a century after the death of Copernicus before the invention of the telescope allowed Galileo to finally disprove Ptolemy's theory of epicycles and show that his model of an earth-centred cosmos was completely wrong. With the telescope Galileo could see that Venus displayed the same full range of phases that the moon did. The diagram below shows that this would have been impossible if the observed path of Venus was due to the planet also following an epicycle, as it orbited between a stationary earth and an earth-orbiting sun, as Ptolemy thought.

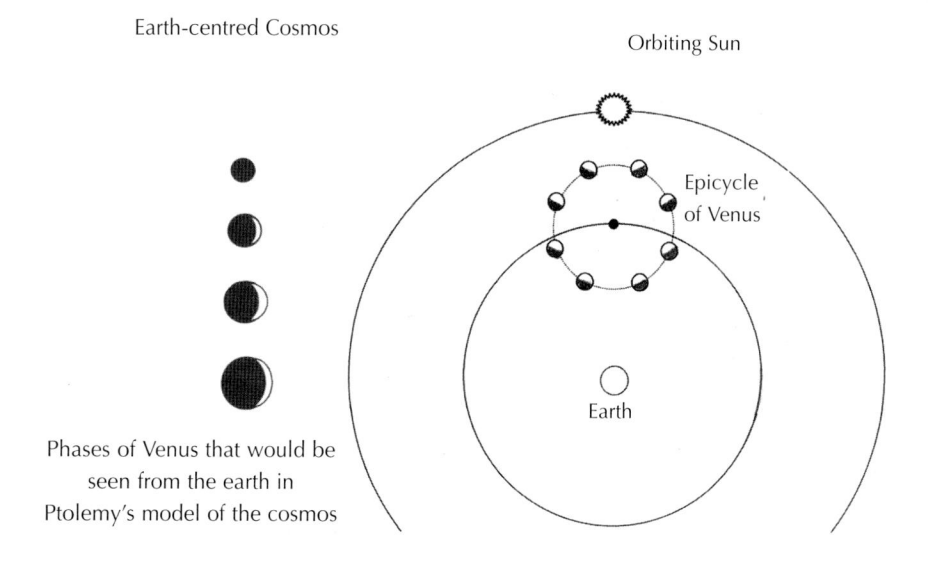

Earth-centred Cosmos

Orbiting Sun

Epicycle of Venus

Earth

Phases of Venus that would be seen from the earth in Ptolemy's model of the cosmos

When they found, or even just suspected, a person of doing or saying anything that might threaten the supreme authority of the Church or its teaching, that person was tortured until they confessed and then executed – normally by being burnt alive at the stake. Even though they knew that they faced a horrible and agonising death, most people were forced by the terrible tortures they had suffered, to confess to being guilty of something simply to escape the awful pain, even if they were genuinely innocent of any crime, or of uttering a blasphemy against the Church. Many of the victims were Jews who had been forced to convert to Christianity. While the Muslims had ruled Spain all religions had been tolerated but the Christian Church in Spain allowed no such freedom and no mercy to unbelievers or doubters.

Unfortunately the influence of the Inquisition did not remain confined to Spain. Gradually over the following centuries the persecution spread to encompass the whole of Roman Catholic Europe. Only those countries that had adopted Protestantism escaped the scourge of torture and execution. Copernicus was fortunate enough to live some time before the more extreme and zealous practises of the Inquisition were to reach the rest of Europe. A later churchman, Canon Giordano Bruno, was not so lucky. He was to suffer persecution and an agonising death for promoting the heliocentric system that Copernicus had proposed.

In fact Bruno's ideas about the universe went far further than even those of Copernicus and were centuries ahead of his time. He put forward the theory that the stars were in reality other incredibly distant suns that may have their own planets orbiting around them, which might be inhabited by other beings. He also even suggested that other churches might exist on these planets with their own version of Christ and the Christian religion. This was too much for the Catholic Church who would not countenance any challenge to their pre-eminence as 'The One True Church' and although Bruno managed to escape their immediate wrath by fleeing into hiding, he was eventually tricked into returning to his native Italy, where after a long imprisonment and torture, like many other victims of the Inquisition, he was burnt alive for his heresy in 1600.

Giordano Bruno's fate in the short-term effectively drove most adherents of the Copernican system underground. One notable exception was the German astronomer, Johannes Kepler (1571–1630). It was Kepler, who in 1601 realised that the orbit of Mars was an ellipse, who was to finally show that the small imperfections in Copernicus's model of the solar system, that had led Copernicus to use the cumbersome device of epicycles, were simply due to the orbits of all the planets being in reality elliptical rather than circular. By doing so he tied up a lot of loose ends that enabled astronomers to explain

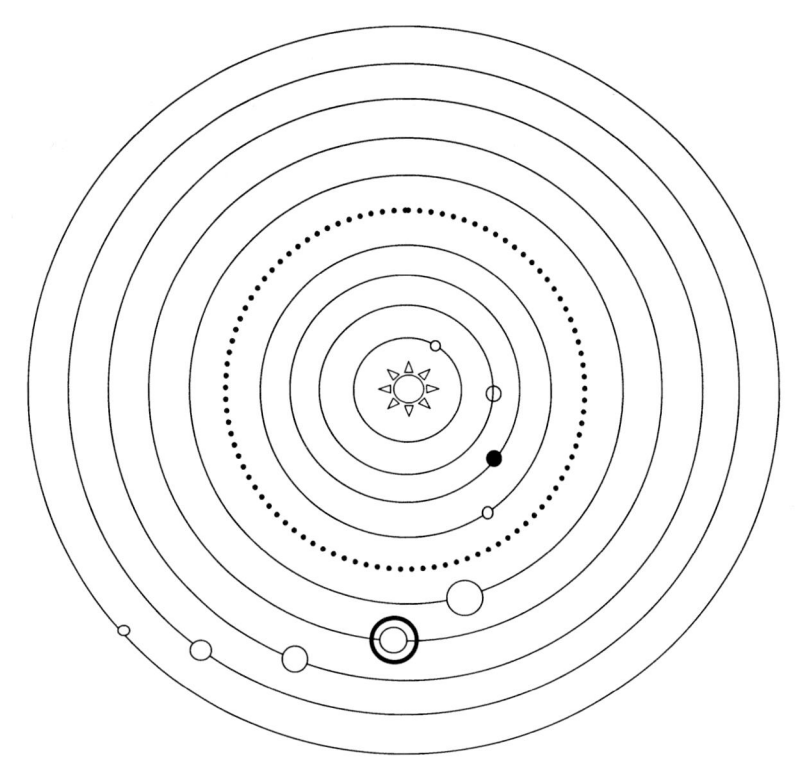

The Solar System

89. There are nine planets in total including the earth that orbit around the sun to form the solar system, with a belt of asteroids between Mars and Jupiter.

Beginning with the sun at the centre they are arranged in the following order: Mercury, Venus, Earth, Mars, the Asteroids, Jupiter, Saturn, Uranus, Neptune, Pluto.

During Copernicus's time the three outermost planets had not been discovered and no-one knew about the belt of asteroids that existed between the orbits of Mars and Jupiter. For a time there was uncertainty about whether Venus or Mercury was nearest to the sun and all that was known about the stars was that they were further away than any object within the solar system, simply because all of these celestial bodies could be seen passing in front of the stars.

some of the previously unaccountable aberrations in their observations. Although he did not know about gravitation, it was also Kepler who suggested that it was magnetic forces that controlled the movements and positions of the planets, paving the way for Isaac Newton.

Despite the support Kepler's work gave to Copernicus's thesis that is not to say there were not doubters even among the most eminent astronomers. Tycho Brahe, for example, whose careful observations were to be used by his student Johannes Kepler to put the final touches to the Copernican model, could not himself accept that the heliocentric system was correct. This was probably in part for religious reasons. Tycho Brahe tried to combine the elements of several theories in his own model of the cosmos, which had each of the other known planets still orbiting the sun but with the sun and moon both orbiting a motionless earth.

Probably the most famous exponent of Copernicus's theories was Galileo Galilei. But although a highly respected figure in his day, even he as an old man was eventually forced by the Church to recant his belief in the heliocentric model in order to save his life. Nevertheless the truth cannot be stifled forever. Most people who gave any thought to the matter recognised the force of Copernicus and Galileo's arguments and not long after the death of Galileo even the Catholic Church had come to reluctantly accept that Copernicus's heliocentric model of the cosmos was the correct one.

Above all else we must remember Copernicus was very much the product of the society that had spawned him. Despite his innovative thinking and willingness to question long-held beliefs, in a culture that far too easily condemned the doubter, he was still in many ways the ultra conservative. He was a canon of the Church, nephew and secretary to a bishop, as well as being his physician and a well-respected mathematician consulted by more than one pope. Unlike the later astronomer Galileo, the doubts he had about Ptolemy's geocentric system he did not broadcast too loudly; nor with any great vehemence and as a result, unlike Galileo he attracted very little unwelcome attention.

What Copernicus did or did not achieve was largely decided by the socio-political conditions and attitudes that prevailed during the period in which he lived. There were many entrenched ideas concerning the earth's place in the cosmos, about the nature and causes of disease, as well as what new technology could be accepted without invoking the charge of sorcery. Even the role of humankind in a universe created by God was subject to a scrutiny that could suppress new thinking or practises. During the Middle Ages there was definitely a socio-political climate in Europe that actually stifled innovative

90. Physical proof of Copernicus's theories of planetary movements was only provided decades after his death, when Galileo first began observing the planets through the newly invented telescope. He provided one particular convincing piece of evidence when he observed phrases of the planet Venus, invisible to the naked eye, these could not be accounted for within the constraints of a geocentric planetary system. He was also the first person to discover that the other planets had moons when he saw the four largest satellites of the planet Jupiter through his telescope. Galileo Galilei was probably the most famous exponent of Copernicus's theories but although a highly respected figure in his day, even he as an old man was eventually forced by the Church to recant his belief in the heliocentric model in order to save his life.

thinking and experiment – even to the extent that intelligent people failed to accept what the evidence of their own eyes and their own rationale was telling them. This occurred at many levels and took many forms.

The attitudes of the established Church were largely responsible for this but the general populace has to bear some of the blame as well. Unquestioning belief in certain religious dogma must have played a large part in perpetuating such a climate of opinion for so many centuries. That people remained for the most part uncritical of the Church's teaching is difficult to understand without recognising how limited the ordinary person's horizon was before the social mobility that resulted due to the depopulation following the Black Death and the growth in social awareness resulting from the invention of the printing press. It is no wonder that the preceding period is so often referred to as the Dark Ages. Unfortunately, this inherent acceptance of the status quo still permeated much of the Middle Ages. In particular it had the effect of discouraging people from pursuing any line of enquiry that might challenge the Church's teaching, or might smack at witchcraft in the eyes of a superstitious people. The penalties for witchcraft or sorcery were sufficiently severe to ensure that few ventured in to uncharted territory when it came to investigating natural phenomena or invention, lest they be charged with sorcery. Moreover, at times, this became practically a political issue: the worse manifestation of this being the invidious Spanish Inquisition.

There were, too, restrictive practices. The extent of these can be best judged by the fact that some clerics even criticised people for wearing spectacles, or for using a fork to eat, because it suggested that the user believed that the God-given attributes they had were not sufficiently effective without employing an artificial aid. This kind of thinking could at times lead to a form of superstitious belief that was in danger of condemning new technology as the work of the devil even though obviously and transparently produced by human agency.

In a more informal way old prejudices had their part to play too: putting a break on forward thinking about such issues as the causes and nature of disease which also discouraged those responsible for public health seeking to implement any really effective preventive measures against the many endemic illnesses that plagued their society. There was a belief held by both the Church and laity, that any malady was simply due to improper living or some form of sin being perpetrated by the sufferer. Not all such beliefs originated from the Church's teaching either. Even ancient pagan rituals were still given credence by both Church and laity – with pre-Christian superstition influencing many walks of life – particularly medicine. It should be remembered that astrology still played an important part in people's lives, being frequently employed by physicians in their diagnosis and was fully accepted by the Christian Church even though its origins went way back to ancient Babylon.

There seems little doubt that in some respects these attitudes discouraged innovation and hindered progress; but there were other more encouraging undercurrents too springing from both internal and external influences, which ultimately affected the medieval world of Europe in a positive way. Warfare as well as trade links bought contact with other cultures and even defeats such as the fall of Constantinople had compensations in the form of the manuscripts containing the ancient wisdom and Eastern knowledge that reached the West. But not only did such events sometimes create the circumstances in which new or rediscovered knowledge became available to European scholars, they also had a wider impact on both society and culture as a whole. Such things had the power to invoke differing perspectives from which new ideas germinated and a new set of values began to emerge alongside the growth in knowledge and technological innovation. These things did not happen overnight but they, nevertheless, helped to promote profound changes in medieval Europe that were to eventually herald the renaissance of knowledge and learning, which in the long-term led to a new order and provided the springboard for social and technological revolution that led to the modern world we know today.

In 1582 Pope Gregory issued an official Church order referred to as a papal bull, instructing and authorising the reformation of the calendar to resolve the problems caused by the inaccuracies in the old Julian Calendar, that had made it very difficult to set the dates of religious festivals like Easter. Two astronomers, Christopher Clavius and Aloysius Lilius, were commissioned to carry out the task. Although Copernicus had never found the time to help Pope Leo III with his own attempts to reform the calendar, he had managed to estimate the length of the sidereal year with remarkable accuracy: to within seconds. It was this figure that the Church referred to when carrying out the later calendar reform for Pope Gregory and it is this version of the calendar that we still use today.

One of the most important things that became evident from Copernicus's new model of the universe was that the earth did not occupy a special place in it; and another was that the recognition that the stars were infinitely distant, which meant that the universe had to be a far bigger place than anyone had previously realised. It was these two innovative concepts that were to change people's thinking and help propel us into the modern world where we began to see the universe as it really was.

The new heliocentric model of the universe proposed by Nicholas Copernicus invoked new concepts in the minds of men and women and provoked new questions that demanded answers. The great Sir Isaac Newton, who finally defined the forces that shape and hold our universe together could not have begun to understand the principles of gravitation if it had not been for Copernicus's new model of the cosmos; and when he said, 'If I have seen further it is because I have stood on the shoulders of giants', one of those giants was undoubtedly Nicholas Copernicus.

GLOSSARY

BIBLIOGRAPHY

A NOTE ON THE ILLUSTRATIONS

INDEX

GLOSSARY

Alexandria	A once-Greek city in Egypt.
Algorismus	About algorisms, which are concerned with the Arabic or decimal system of counting.
Allenstein	Polish town.
Apparent	Movement, size, etc. Perceived rather than actual phenomenon.
Arabic numerals	Modern numerals and the earlier Arabic-Indian forms upon which they were based.
Armillary sphere	An astronomical instrument indicating the apparent path and position of the stars, planets, moon and sun.
Astrolabe	An astronomical instrument.
Astronomical unit	Distance between earth and sun.
Astronomy	Study of the cosmos.
Baltic Sea	Between Eastern Europe and Asia.
Bologna	Italian city.
Bruges	Belgian city.
Camera lucida	A device that superimposes an object's image on for example an artist's drawing or painting surface.
Camera obscura	A darkened room with a pinhole aperture allowing external images to be cast on a reflective surface.
Canon law	Church law.
Canon	An administrative, clerical post in the Church.
Celestial clock	The nightly, predictable, apparent movements of the stars.
Celestial globes	A globe with the positions of the stars marked upon it.

Chaldean astronomy	The astronomy originating in ancient Mesopotamia.
Clergy	Those of the Church.
Cologne	German city.
Compass	A device for determining the direction of north.
Constantinople	The now Turkish city of Instanbul.
Copernican	Pertaining to Copernicus.
Cordova	Spanish city.
Cross-staff	A device using trigonometry for determining angles and distances.
Danzig	Polish city.
Deferent	An orbit around a non-central body.
Domine praeceptor	Latin for learned teacher.
Duchy	Region ruled by either a duke or duchess or one of equivalent rank.
Eclipse (Lunar)	When the earth's shadow either partially or totally covers the moon.
Eclipse (Solar)	When the moon moves between the earth and the sun obscuring its light.
Ecliptic	The apparent path of the sun across the sky.
Elevation	The height or angle of an object above the horizon.
Ephemerides	A set of tables to aid navigation by predicting the position of heavenly bodies at various times and at different locations.
Epicycles	A smaller circular movement of a body in addition to its orbital motion.
Equator	An imaginary line dividing a planet's northern and southern hemispheres.
Equinoxes	The time of year when day and night are of equal length and the sun rises and sets due east and west (in the northern hemisphere).
Ermland	Region of Prussia.
Ferrara	Italian city.
Flanders	Belgian city.
Florence	Italian city.
Frauenburg/Fromburg	Polish city.

GLOSSARY

Hanseatic League	A medieval association or guild of Germanic merchants and traders.
Heilsburg/Heilsberg	Polish city.
Heliocentric	Sun-centred.
Hemlock	Highly poisonous plant.
Herbalists	Those using plants for medicinal purposes.
Heresy	Statements or practises challenging the teaching of the Church.
Hippocratic oath	A promise traditionally made by doctors based upon the principles defined by the ancient Greek physician Hippocrates.
Hockney, David	Well-known, present-day artist.
Holy Roman Empire	A medieval alliance of mainly Germanic states.
Hospitallers	Knights dedicated to treating those ill or injured.
Humanists	Those promoting the human virtues.
Hyoscyamus	A medicinal plant.
Inclination	Angle at which the earth's axis leans towards or away from the sun.
Inferior planet	Planet with an orbit inside the earth's orbit.
Inquisition	A group of clerics appointed by the pope to seek out and punish heretics.
Jubilee	A centenary celebration.
Julian Calendar	Calendar revised under the auspices of Julius Caesar.
Jupiter	Fifth planet from the sun.
Kraków	Polish city.
Kulm	Polish city.
Latin	Language of ancient Rome.
Latitude	Distance north or south of the Equator.
Leipzig	Polish city.
Leprosy	Infectious disease of the nervous system resulting in damage to the body's extremities due to loss of feeling.
Literacy	The ability to read.
Lunar	Pertaining to the moon.
Lutheran	Practices pertaining to the teachings of Martin Luther.
Machiavellian	Manipulative.

GLOSSARY

Mars	Fourth planet from the sun.
Retrograde movement	An apparent reverse in a planet's movement.
Mercury	The planet nearest to our sun.
Moon	The satellite of a planet.
Saturn	The sixth planet from our sun.
Navigation	The art or science of establishing one's location and direction of travel.
Nocturne	Device for telling the time at night from the position of the stars.
Node	A points where there is an intersection with the plane of the Ecliptic (coinciding with the apparent path of the sun).
Novgorod	Russian city.
Nuremberg	German city once capital of the Holy Roman Empire.
Occultation	The passing of one body in front of another.
Opium	Narcotic produced from a poppy.
Optics	A study of light, visual phenomena and perception.
Orbit	The path of an object circling around another body.
Padua	Italian city.
Papal	Pertaining to the pope.
Parallax	A method of using trigonometry to estimate the distances of objects.
Phase	One of a series of changes such as when the surface of a body is observed to be illuminated to varying amounts by the sun.
Physicians	Medical doctor.
Plague	Black Death.
Planetary systems	Arrangement of orbiting planets.
Planets	A body orbiting a sun.
Precession	Gradual cyclic advancement.
Ptolemaic	Pertaining to Ptolemy.
Ptolemy's rulers	A device to measure the elevation of an object above the horizon (see Triquetrum).
Pythagoran	Pertaining to Pythagoras.
Pythagorans	The followers of Pythagoras.
Regression/retrogression	Reversal or going backwards.

GLOSSARY

Retrograde movement	(see regression).
Riga	Polish city.
Roman numerals	Numerals based on letters as used in ancient Rome.
Rotation	A body turning on its own axis.
Samos	A Greek island.
Saturn	Sixth planet from the sun.
Sidereal	Relating to the stars.
Sidereal time	Time based on the apparent movement of the stars.
Silesia	A region of Germany.
Sorcery	The use of spells and magic.
Star	A distant sun.
Sundial	A device for telling the time from the position of a shadow cast by the sun.
Syene	A city in Asia Minor.
Syphilis	One of several sexually transmitted disease.
Terrestrial globe	A globe showing the features of the earth.
Teutonic Knights	An order of Germanic knights.
Torun	Polish city that was the birthplace of Nicholas Copernicus.
Transit	The passing of one heavenly body across the surface of another such as the sun being passed by an inferior planet.
Trigonometry	A method using the properties of triangles to measure distances and dimensions.
Triquetrum	(see Ptolemy's rulers).
Tropic of Cancer	The tropic north of the Equator.
Tropic of Capricorn	The tropic south of the Equator.
Tropics	Two imaginary circles defining the points at which the sun appears directly overhead at the Summer Solstices in the northern hemisphere and the southern hemisphere.
Universe	Everything that exists.
Vatican	Official seat of the pope in Rome.
Venus	Second planet from the sun.
Wittenberg	A German city.

BIBILOGRAPHY

Adamczewski, Jan, 1972, *Nicolaus Copernicus and his Epoch* (Interpress Publishers, Warsaw)

Armatige, Angus, 1951/47, *The World of Copernicus* (*Sun stand thou still*) (Ep Publishing Ltd, England)

Chapman, Allan, 2001, *Gods in the Sky* (Channel 4)

De Pree, C. and Axelrod, A., 1999, *The complete idiots guide to astronomy* (Alpha Books)

Hartmann, William (ed.), 1991, *Astronomy: the cosmic journey* (Wadsworth Publishing Company, CA)

Holmes, George (ed.), 2001, *The Oxford illustrated history of Medieval Europe* (OUP)

Hoskin, Michael (ed.), 1997, *Cambridge Illustrated History of Astronomy* (CUP)

Hoskin, Michael (ed.), 1999, *The Cambridge Concise History of Astronomy* (CUP)

Kesten, 1946, *Copernicus and his World* (Secker And Warburg, London)

Kolb, Rocky, 1999, *Blind Watchers of the Sky* (OUP)

Kuhn, Thomas S., 1966, *The Copernican revolution* (Harvard University Press, Cambridge)

Le Goff, Jacques (ed.), 1990, *The Medieval World* (Collins and Brown Ltd, London)

Levy, David, 1995, *Skywatching: the ultimate guide to the Universe* (Harper Collins Publishers, London)

Mitton, Jacqueline, 1991 *A concise dictionary of astronomy* (OUP)

Mizwa, Steven P., 1969, *Nicholas Copernicus 1543–1943* (Kennikat Press Inc., New York)

North, John, 1994, *Astronomy and cosmology* (Fontana Press/Harper Collins Publishers, London)

Peterson, Ivars, 1993, *Newton's Clock: chaos in the solar system* (W.H. Freeman and Company, NY)

Sinclair, Rohde, 1974, *The Old English Herbal* (Minerva Press Ltd, London)

A NOTE ON THE ILLUSTRATIONS

The diagrams, maps and several of the illustrations in this book have been produced by me, the author; but in order to ensure authenticity many of the other illustrations have been based upon contemporary medieval or Renaissance images. The reader should be aware that the majority of these have been redrawn or adapted often with some modification to the composition and therefore depart to a degree from the original. The artists that first created these images are all long dead but in some instances the illustrations in this book have been based upon information gleaned from reproductions. In the majority of cases every effort has been made to ensure that the new images produced are original works of art and are, therefore, not subject to copyright held by any other party. Several, although redrawn, still remain close to the original and a few are unmodified. Those already in the public domain are not subject to copyright. In other cases every effort has been made to establish who owns the copyright but in some cases this has proved unsuccessful. No infringement of copyright has been intended but should anyone feel this has inadvertently occurred, please contact the author in the first instance via the following website: www.pastworld.net

Introduction
1. Map of medieval Europe

Chapter 1
2. The town of Toruń
3. Brutal melee
4. Jewish moneylender
5. Copernicus's house
6. Medieval merchants

Chapter 2
7. Printing press
8. Paper making
9. Scribe
10. Spectacles
11. Pope Gregory IX
12. Setting up a sundial
13. Celestial Clock
14. Telling the time at night

Chapter 3
15. Fall of Constantinople
16. Islamic Academy
17. Roman and Arabic numerals

Chapter 4
18. Ermland and Teutonic Knights
19. Kraków
20. Education
21. Cross-staff
22. Ptolemy

Chapter 5
23. The astrolabe
24. Using an astrolabe
25. Armillary sphere
26. Determining latitude at night
27. Using a cross-staff to find latitude
28. Using a quaadrant
29. Bishop Lucas Waczenrode
30. Estimating distance of a celestial object
31. Lecture in mathematics
32. The Borgias
33. Nicholo Machiavelli

Chapter 7
34. A dissection
35. University Students
36. Medieval physician resetting limbs
37. Poisonous plants used as anaesthetics

LIST OF ILLUSTRATIONS

Chapter 8
38. Health and astrology
39. Heilsberg
40. Teutonic Kinght

Chapter 9
41. The tilted earth
42. Inclination of the earth to the eliptic
43. How Erastothenes estimated the size of the earth
44. Determining latitude from the position of the sun
45. Estimating the size of the moon
46. Calculating the distance of the moon
47. Calculating the distance of the sun from the earth
48. Estimating the size of the sun
49. An explanation of Ptolemy's thinking
50. The retrograde path of Mars
51. Epicycles
52. The movement of the planets
53. An inferior planet at greatest elongation
54. Retrograde movements of Mars
55. Retrograde planetary motion
56. Frauenburg
57. Leprosy
58. Nicholas Copernicus
59. Voyages of Columbus
60. Treating the sick
61. Prayer to the earht goddess
62. The plague
63. Herb garden
64. Herbal remedies

Chapter 11
65. Pope Leo X
66. The precession of the equinoxes
67. Lunar precession

Chapter 12
68. Leonardo da Vinci
69. The granting of Indulgences
70. Martin Luther

Chapter 13
71. Copernicus's tower at Frauenburg
72. Pope Innocent III
73. Allenstein Castle
74. Medieval soldiers looting a castle
75. A treatise on coinage

Chapter 14
76. Tiedemann Giese
77. Bishop Dantiscus

Chapter 15
78. Copernicus suing a triquetrum
79. Pope Clement VII
80. Parallax of the stars

Chapter 16
81. Nuremburg at the time of Copernicus
82. Comparing the size of the orbits of the earth and other planets
83. The conditions required to see a transit of Venus
84. Andrew Vesalius
85. Death of Copernicus

Chapter 17
86. Canon Bruno being burnt at the stake
87. Johannes Keplar
88. Phases of Venus
89. The solar system
90. Galileo

INDEX by topic

PERSONALITIES

Albert, Grand Master of the Teutonic
    Knights 131, 133, 147
Albert of Prussia (Duke) 147 see above
Andrew II of Hungary 13,
Archimedes 35, 83, 92, 120
Aristarchus of Samos (astronomer) 83, 87,
    89, 90, 120–121, 163
Armatus, Salvinus de 26

Bacon, Roger (Friar) 26–29
Bolingbroke, Henry 17
Borgia, Cesare 60, 62, 63
Borgia, Lucrezia 60, 62
Borgia, Rodrigo (Cardinal) 60, 62, 63
Borgognoni, Theodoric 69
Brahe, Tycho (astronomer) 170
Brudzewski, Albert 43
Bruno, Giordano (Canon) 160, 168
Bylica, Marcia (astronomer) 49

Canaletto (artist) 30
Casimir the Great (Polish king) 42
Catanei, Vannozza 60
Celibacy 138–139
Charles of Burgundy 17
Christopher Clavius 173
Commissioner of Ermland 132, 134, 137
Copernicus, Andreas 42, 56, 59, 60, 61,
    102–105

Danticus (Bishop) 138–139, 145
Donner, George (Canon) 151
Duke of the Romagna 60

Eratosthenes 81–82, 85, 87

Ferber (Bishop) 137
Fibonacci, Leonardo (mathematician) 38

Galen 66–67
Galileo (astronomer) 167, 170–171
Galioni, Alesandra 66
Giese, Tiedamann (Canon/Bishop)
    136–139, 146–147, 151, 154, 158
Goethal, Heindrich 26
Gresham, Sir Thomas 135
Gutenburg, Johann 22, 23

Halley, Edmund (astronomer) 153, 163
Harrison (clockmaker) 119
Henry IV of England 17
Hipparchus 83, 88, 89, 119, 163
Holy Roman Emperor 132, 134
Hypocrites 67
Hockney, David 30
Holywood, John 44

Ingres (artist) 30

Julius Caesar 115

Keplar, Johannes (astronomer) 146, 162,
    164, 168, 170
Kohler 144
Konrad of Masovia (Duke) 12, 15, 16, 129

Lilius, Aloysius 173
Lippershey, Hans 29
Lossainen, Fabian (Bishop) 137
Lucca, Hugh of 69
Lucretius, Cognominatus 142-143
Luther, Martin 124-127, 136, 145, 146

Machiavelli 61, 63
Mueller, Johann 44
Mundinus of Bologne 66
Muschenbroeck, Peter Von 26

INDEX

Newton, Isaac (Sir) 170, 174
Novara, Domenico Maria Da 59, 60, 73, 83, 148

Osiander, Andreas 154, 154, 157

Peuerbach, Georg 45
Popes:
  Alexander VI 61, 62, 63
  Clement VII 142, 143
  Gregory IX 29, 173
  Innocent III 129, 130
  Leo X 114, 117, 154, 173
  Martin IV 25
Ptolemy (astronomer) 35, 45, 46, 47, 60, 91, 92, 93, 96. 97, 120, 140, 141, 143, 144, 149, 167

Rabe (friend) 78
Regiomuntanus (astronomer) 44, 45, 46, 120
Rivalto di, Giordano (Friar) 25, 27

Rheticus, George 61, 145–147, 151, 154, 155, 158–159

Sacrobosco, John 44, 46
Saliceto, Guglielmo de (William of Saliceto) 67, 69
Schilling, Anne (mistress) 138–139, 151
Schoener, Johannes 147
Schonberg (Cardinal) 142–143, 147
Sculteti, Alexander (Canon) 138–139, 151
Sigismund (king) 101, 131
Spina, Alessandro della (Friar) 25, 26

Torre, Marcus Antonius de la 67

Vesalius, Andreas 154
Vinci, Leonardo da 67, 122–124

Waczenrode, Barbara 21
Waczenrode, Lucas (Bishop) 41, 42, 56, 57, 59, 60, 77–79, 100, 101, 158
Wurtzburg, John 13

GROUPS

Arabs 24, 37, 57

Canons 57
Chinese 25
Clergy, Clerics 26, 172

Entrepreneurs 18

Franciscans 27

Goths 35

Hanseatic League 18, 20
Herbalists 69–70, 106, 109, 111–113
Hospitallers 13, 129
Humanists 124, 125

Immigrants 20

Jews 18, 24, 36, 168

Lithuanians 16, 17, 42
Lutherans 145

Mercenaries 16
Merchants 17, 18, 20, 21
Muslims 13, 24, 25, 36, 37, 42, 157, 168

Normans 37, 157

Pagans 16
Physicians 65–69, 104–109
Prussians 12–18
Pythagorans 146

Teutonic Knights (Order of) 12–18, 42, 76–78, 100, 128, 129–134
Tradesmen and tradeswomen 18
Turks 34, 35

Vandals 35

INDEX

SCIENCE, CULTURE AND TERMINOLOGY

Abacus (book of) 38
Absolution 129
Algebra 45
Anaesthetics and soporifics 69–70
Arabic/Indian numerals 37–39, 157
Astrology 72, 73, 75

Calendar reform 114–120, 173
Camera lucida 30
Camera obscure 30
Canon law 59, 63, 75
Clocks 31, 33, 46, 82
Commissioner of Ermland 132, 134, 137
Copying by hand 26
Cross-staff 45, 46, 53, 135, 141
Crusades 37

Dark Ages 81, 83, 172
Disease 101–105, 107–108, 110
Dissection 64–67

Education 24, 35, 41–44, 75
Epact 117
Execution 61, 64, 65

Firearms 133

Geocentric System 47
Glass-making 24
Greek/Greek manuscripts 35, 38, 43, 75, 121
Gresham's law 135

Herbal remedies 69–70. 106, 109, 111–113
Heresy 127
Hippocratic oath 67
Homosexuality 145, 159

Indulgencies 125
Inquisition 18, 29, 124, 130, 172

Jubilee 60, 62
Julian Calendar 114–115

Latin 21, 43, 44, 45, 75

Lenses 26, 29
Leprosy 102–104
Literacy 22, 23, 59

Map-making 55, 75
Mathematics 35–39, 61
Medicine 35, 38, 64, 65–73, 102–113, 172
Microscope 24

Navigation 48, 54
Nocturnal 33
Node 153
Numeracy 59
Numerals:
    Arabic/Indian 37–39, 157
    Roman 36-38

Optics 24–31

Pagan rituals/beliefs 109, 112–113
Papal exemption 130
Paper 24, 25
Parchment 24
Plague 107–108, 110
Planets known:
    Mercury 85, 93
    Venus 85, 93, 163
    Mars 91, 93, 97, 165, 168
    Jupiter 91, 93, 97
    Saturn 91, 93, 97
Printing 22, 24
Physicians 65–69, 104–109
Publications mentioned:
    Almagest 45
    Algorismus 46
    Book of the Abacus 38
    Ephemerides 46
    First Account of the Book of Revolutions
    Little Commentary 79, 142
    Narratio Prima de Libris Revolutionum 147, 151
    De Revolutionibus Orbium Coelestium Libri VI 150, 151, 155–158
    (*Six Books concerning the Revolution of the Spheres*)

On the sizes & distances of the sun & the moon 83, 89
Preface: original/substitute 154–157, 166
The Sleepwalkers, Kohler 165
The Prince 61
The Sphere 46
Treatise on Coinage 134, 135

Quadrant 52, 54

Reformation 124–127, 136
Roman numerals 36–37

Secret knowledge 30
Serfdom 130
Sextant 50
Siege 133
Solar System 79

Sorcery 27, 29, 30, 172
Spectacles 25–28
Sundials 31–33, 45
Surgery 68, 69
Surveyors 45, 48, 55
Syphilis 103–105

Telescope 24, 28, 29, 91, 141, 163, 167
Timekeeping 31–33, 46
Trigonometry 135, 148

Vatican 124
Vellum 24

Warfare 13–18, 129–134

Zero 38, 39
Zodiac 75

ASTRONOMY

Aldebaran (star) 92
Algorismus 46
Alignment 80
Altitude 48
Armillary sphere 51, 142
Astrolabe 48–51, 135
Astrology 72, 73, 75
Astronomical unit 163

Babylonian/Chaldean/Mesopotamian astronomy 75, 119, 144, 173

Calendar reform 114–120
Celestial Clock 32, 33
Compass
Cross-staff 45, 46, 53, 135

Deferent 97
Distances of:
  Moon 58, 59, 83, 88
  Planets 58, 152, 153, 163
  Stars 91, 148, 149, 168
  Sun 58, 89, 153, 163

Earthlight 122–124
Eclipse 83, 87, 88, 91
Ecliptic 53, 92

Elevation 48
Elliptical orbits 144–146, 162, 164
Elongation 94–95, 97, 152
Epact 117
Ephemerides 46
Epicycles 93, 96–97, 144–146, 162, 164
Equator 55, 82
Equinoxes 82, 116–120

Geocentric planetary model theory 47, 59, 91, 144, 162
Globes:
  celestial 53
  terrestrial 53

Heliocentric planetary model/theory 79, 120, 149, 163

Inclination 80–86

Julian Calendar 114–115
Jupiter (planet) 80, 91

Latitude 46, 52, 55
Longitude 46
Lunar eclipse 61, 82
Lunar time 119

INDEX

Mars (planet) 80. 91, 92, 93, 94, 96
Mercury (planet) 80, 91, 93, 95
Moon 58, 59, 83, 87, 88, 89
Moons of Jupiter 171

Navigation 45, 46, 48, 52–55
Nocturnal 32, 33

Occultation 91, 92
Orbits 80, 93, 94, 95

Parallax 58, 59, 83, 148–149, 152, 163
Planetary systems 47, 59, 79, 91, 169
Planets 80, 91, 92, 93, 152, 163, 169
Pole Star (*Polaris*) 31, 32, 33, 52, 53, 55, 116
Precession of the equinoxes 116–120
Ptolemy's rulers 53, 140, 141
Publications mentioned: on mathematics and astronomy:
   Almagest 45
   Algorismus 46
   Book of the Abacus 38
   Ephemerides 46
   First Account of the Book of Revolutions 147
   Little Commentary 79, 142
   Narratio Prima de Libris Revolutionum 147, 151
   De Revolutionibus Orbium Coelestium Libri VI 150, 151, 155–158
   (*Six Books concerning the Revolution of the Spheres*)
   On the sizes & distances of the sun and the moon 83, 89
   Preface: original/substitute 154–157, 166
   The Sphere 46
   The Sleepwalkers, Kohler 165
Pythagorans 146

Quadrant 52, 54

Regression 92, 96–99
Retrograde movement 92, 96–99, 144
Rotation 53, 80, 82, 165

Saturn (planet) 80, 91, 92
Seasons 81–84, 86, 99
Sidereal time/year 32–33, 116, 120, 173
Sizes of:
   Earth 81–82
   Moon 87, 90
   Planetary orbits 152, 153, 163
   Sun 90
   Universe/Cosmos 58, 149, 173
   Solar System 58, 80, 152, 163, 169
Solar System 58, 80, 152, 169
Solar time/year 116
Summer Solstice 82, 117
Sun 55, 58, 89, 90, 153, 163
Sundial 31, 32, 33, 53, 142

Telescope 24, 28, 29, 91, 141, 163, 167
Timekeeping 31–33
Trigonometry 58, 152
Triquetrum 49, 51, 53, 140, 141
Tropics 50, 81–86

Ursa Major 33

Venus (planet) 80, 91, 93, 95
   transit 153, 163
   phases 167

Winter Solstice 82

Zodiac 53, 75

GEOGRAPHICAL LOCATIONS

Allenstein Castle 131, 133
Alexandria 35, 38, 47, 81, 83, 119

Baltic 20
Bologna 59, 60, 62, 66, 69–71, 75, 83, 142, 148

Bruges 20

China 20, 22, 24, 36
Cologne 20
Constantinople 34, 35, 46, 157, 173
Cordova 36

GEOGRAPHICAL LOCATIONS, CONTINUED

Danzig 20

Egypt 24, 35, 38, 75, 81, 157
England 142
Ermland 17, 40, 42, 77, 100, 129–134, 145

Ferrara 62, 63
Flanders 20
Florence 25
Fraunenburg 57, 59, 61, 62, 100–103, 128, 130, 131, 133, 137–139, 142

Germany 16, 20, 130
Greece 35, 38, 42, 57, 67, 157

Heilsberg 74, 77–79, 101
Holy Roman Empire 150
Hungry 13, 14, 20

Kraków 21, 42, 43, 57, 101, 141
Kulm 139

Leipzig 151
Lithuania 17, 135
Loebau Castle 146–147

Mainz 23
Mediterranean Sea 24
Mesopotamia 75

Novgorod 20
Nuremberg 150, 154

Padua 62, 65–71
Poland 12–21, 42, 77, 129, 135
Prussia 12–18, 20, 129

Rhineland 20
Riga 20
Rome (ancient) 24, 35, 42, 46, 57, 81, 83, 157
Rome (medieval) 35, 60, 61, 62
Russia 17, 20

Sicily 37, 42, 157
Silesia 20
Spain 24, 25, 36, 37, 42
Syene 81, 82, 83

Tannenberg (battle of) 18, 130
Torun 12, 18, 20

Wittenberg 124, 145